CAMBRIDGE TRACTS IN MATHEMATICS

General Editors

234 The Theory of Countable Borel Equivalence Relations

CAMBRIDGE TRACTS IN MATHEMATICS

A complete list of books in the series can be found at www.cambridge.org/mathematics.
Recent titles include the following:

200. Singularities of the Minimal Model Program. By J. Kollár
201. Coherence in Three-Dimensional Category Theory. By N. Gurski
202. Canonical Ramsey Theory on Polish Spaces. By V. Kanovei, M. Sabok, and J. Zapletal
203. A Primer on the Dirichlet Space. By O. El-Fallah, K. Kellay, J. Mashreghi, and T. Ransford
204. Group Cohomology and Algebraic Cycles. By B. Totaro
205. Ridge Functions. By A. Pinkus
206. Probability on Real Lie Algebras. By U. Franz and N. Privault
207. Auxiliary Polynomials in Number Theory. By D. Masser
208. Representations of Elementary Abelian p-Groups and Vector Bundles. By D. J. Benson
209. Non-homogeneous Random Walks. By M. Menshikov, S. Popov, and A. Wade
210. Fourier Integrals in Classical Analysis (Second Edition). By C. D. Sogge
211. Eigenvalues, Multiplicities and Graphs. By C. R. Johnson and C. M. Saiago
212. Applications of Diophantine Approximation to Integral Points and Transcendence. By P. Corvaja and U. Zannier
213. Variations on a Theme of Borel. By S. Weinberger
214. The Mathieu Groups. By A. A. Ivanov
215. Slenderness I: Foundations. By R. Dimitric
216. Justification Logic. By S. Artemov and M. Fitting
217. Defocusing Nonlinear Schrödinger Equations. By B. Dodson
218. The Random Matrix Theory of the Classical Compact Groups. By E. S. Meckes
219. Operator Analysis. By J. Agler, J. E. McCarthy, and N. J. Young
220. Lectures on Contact 3-Manifolds, Holomorphic Curves and Intersection Theory. By C. Wendl
221. Matrix Positivity. By C. R. Johnson, R. L. Smith, and M. J. Tsatsomeros
222. Assouad Dimension and Fractal Geometry. By J. M. Fraser
223. Coarse Geometry of Topological Groups. By C. Rosendal
224. Attractors of Hamiltonian Nonlinear Partial Differential Equations. By A. Komech and E. Kopylova
225. Noncommutative Function-Theoretic Operator Function and Applications. By J. A. Ball and V. Bolotnikov
226. The Mordell Conjecture. By A. Moriwaki, H. Ikoma, and S. Kawaguchi
227. Transcendence and Linear Relations of 1-Periods. By A. Huber and G. Wüstholz
228. Point-Counting and the Zilber–Pink Conjecture. By J. Pila
229. Large Deviations for Markov Chains. By A. D. De Acosta
230. Fractional Sobolev Spaces and Inequalities. By D. E. Edmunds and W. D. Evans
231. Families of Varieties of General Type. By J. Kollár
232. The Art of Working with the Mathieu Group M_{24}. By R. T. Curtis
233. Category and Measure. By N. H. Bingham and A. J. Ostaszewski
234. The Theory of Countable Borel Equivalence Relations. By A. S. Kechris

The Theory of Countable Borel Equivalence Relations

ALEXANDER S. KECHRIS
California Institute of Technology

Shaftesbury Road, Cambridge CB2 8EA, United Kingdom

One Liberty Plaza, 20th Floor, New York, NY 10006, USA

477 Williamstown Road, Port Melbourne, VIC 3207, Australia

314–321, 3rd Floor, Plot 3, Splendor Forum, Jasola District Centre,
New Delhi – 110025, India

103 Penang Road, #05–06/07, Visioncrest Commercial, Singapore 238467

Cambridge University Press is part of Cambridge University Press & Assessment,
a department of the University of Cambridge.

We share the University's mission to contribute to society through the pursuit of
education, learning and research at the highest international levels of excellence.

www.cambridge.org
Information on this title: www.cambridge.org/9781009562294
DOI: 10.1017/9781009562256

When citing this work, please include a reference to the DOI 10.1017/9781009562256

First published 2025

A catalogue record for this publication is available from the British Library

*A Cataloging-in-Publication data record for this book is available from
the Library of Congress*

ISBN 978-1-009-56229-4 Hardback

To Olympia, Sotiri, Katerina, David, Alexi and Bia

Contents

Preface

The theory of definable equivalence relations has been a very active area of research in descriptive set theory over the past three decades. It serves as a foundation of a theory of complexity of classification problems in mathematics. Such problems can often be represented by a definable (usually Borel or analytic) equivalence relation on a standard Borel space. To compare the difficulty of one classification problem with respect to another, one introduces a basic order among equivalence relations, **Borel reducibility**, which is defined as follows. An equivalence relation E on a standard Borel space X is Borel reducible to an equivalence relation F on a standard Borel space Y, in symbols $E \leq_B F$, if there is a Borel map $f\colon X \to Y$ such that $xEy \iff f(x)Ff(y)$. In this case one views E as less complex than F. The study of this hierarchical order and the discovery of various canonical benchmarks in this hierarchy occupies a major part of this theory.

Another source of motivation for the theory of definable equivalence relations comes from the study of group actions, in a descriptive, topological or measure-theoretic context, where one naturally studies the structure of the equivalence relation whose classes are the orbits of the action and the associated orbit space.

An important part of this theory is concerned with the structure of **countable Borel equivalence relations**, i.e., those Borel equivalence relations all of whose classes are countable. It turns out that these are exactly the equivalence relations generated by Borel actions of countable discrete groups (Feldman–Moore), and this brings into this subject important connections with group theory, dynamical systems and operator algebras.

Our goal here is to provide a survey of the state of the art in the theory of countable Borel equivalence relations. Although this subject has a long history in the context of ergodic theory and operator algebras, the systematic study of countable Borel equivalence relations in the purely Borel context dates back to the mid-1990s and originates in the papers [DJK] and [JKL]. Since that time there has been extensive work in this area leading to major progress on many of the fundamental problems.

This survey is organized as follows. Chapter 1 reviews some basic general concepts concerning equivalence relations and morphisms between them. In Chapter 2 we introduce countable Borel equivalence relations, discuss some of their properties and mention several examples. The scope of the theory of countable Borel equivalence relations is actually much wider as it encompasses a great variety of other equivalence relations up to Borel bireducibility. Chapter 3 deals with such equivalence relations, called **essentially countable**. In Chapter 4, we consider invariant as well as quasi-invariant measures for equivalence relations.

In Chapter 5, we start studying the hierarchical order of Borel reducibility, introducing the important benchmarks of the simplest (nontrivial) and the most complex countable Borel equivalence relations. The next chapter (Chapter 6) demonstrates the complexity and richness of the structure of this hierarchical order and discusses the role of rigidity phenomena in both the set-theoretic and the ergodic-theoretic contexts. We next consider various important classes of countable Borel equivalence relations: hyperfinite (Chapter 7), amenable (Chapter 8), treeable (Chapter 9), freely generated (Chapter 10) and finally universal ones (Chapter 11). The next four chapters deal with the algebraic structure of the Borel reducibility order (Chapter 12), the concept of structurability of countable Borel equivalence relations (Chapter 13), topological realizations of countable Borel equivalence relations (Chapter 14) and a universal space for actions and equivalence relations (Chapter 15). The final chapter (Chapter 16) collects many of the main open problems discussed earlier.

With a few minor exceptions, this survey contains no proofs, but it does include detailed references to the literature where these can be found. The emphasis here is primarily on the descriptive aspects of the theory of countable Borel equivalence relations, and ergodic-theoretic aspects are brought in when relevant. It is not our intention, however, to survey the research in ergodic theory related to this subject, including in particular the theory of orbit equivalence and its relation to operator algebras. This can be found, for example, in [Z2], [AP] and [I3], the last of which also contains a detailed bibliography of the extensive work in this area over the past two decades. Exposition of other related subjects, such as the Levitt–Gaboriau theory of cost, can be found in [KM1]. Finally there are important connections with descriptive aspects of graph combinatorics, for which we refer the reader to [KM] and [Pi].

Acknowledgments. The author was partially supported by NSF Grants DMS-1464475 and DMS-1950475.

I would like to thank R. Chen, C. Conley, J. Frisch, A. Ioana, A. Marks, C. Meehan, S. Mejak, B. Miller, C. Rosendal, F. Shinko, S. Thomas, A. Tserunyan, R. Tucker-Drob and M. Wolman for many helpful comments and suggestions.

1

Equivalence Relations and Reductions

1.1 Generalities on Equivalence Relations

Let E be an equivalence relation on a set X. If $A \subseteq X$, we let $E \upharpoonright A = E \cap A^2$ be its **restriction** to A. We also let $[A]_E = \{x \in X : \exists y \in A(xEy)\}$ be its **E-saturation**. The set A is **E-invariant** if $A = [A]_E$. In particular, for each $x \in X$, $[x]_E$ is the **equivalence class**, or **E-class**, of x. A function $f : X \to Y$ is **E-invariant** if $xEy \implies f(x) = f(y)$. Finally, $X/E = \{[x]_E : x \in X\}$ is the **quotient space** of X modulo E.

Suppose that E, F are equivalence relations on sets X, Y, respectively, and $f : (X/E)^n \to Y/F, n \geq 1$, is a function. A **lifting** of f is a function $\tilde{f} : X^n \to Y$ such that $f(([x_i]_E)_{i<n}) = [\tilde{f}((x_i)_{i<n})]_F, \forall x \in X$. Similarly if $R \subseteq (X/E)^n$, its lifting is $\tilde{R} \subseteq X^n$, where $(x_i)_{i<n} \in \tilde{R} \iff ([x_i]_E)_{i<n} \in R$.

If E_i, $i \in I$, is a family of equivalence relations, with E_i living on X_i, we define the **direct sum** $\bigoplus_i E_i$ to be the equivalence relation on $\bigoplus_i X_i = \{(x, i) : x \in X_i\}$ defined by

$$(x, j) \bigoplus_i E_i \ (y, k) \iff j = k \ \& \ xE_j y.$$

In particular, we let for $n \geq 1$, $nE = \bigoplus_{i<n} E$. Also let $\mathbb{N}E = \bigoplus_{i \in \mathbb{N}} E$.

We define the **direct product** $\prod_i E_i$ to be the equivalence relation on the space $\prod_i X_i$ defined by

$$(x_j) \prod_i E_i \ (y_j) \iff \forall j(x_j E_j y_j).$$

In particular, we let for $n \geq 1$, $E^n = \prod_{i<n} E$. Also let $E^{\mathbb{N}} = \prod_{i \in \mathbb{N}} E$.

If E, F are equivalence relations on X, then $E \subseteq F$ means that E is a subset of F, when these are viewed as subsets of X^2, i.e., E is finer than F or equivalently F is coarser than E. The **index** of F over E, in symbols $[F : E]$, is the supremum of the cardinalities of the sets of E-classes contained in an

F-class. Thus $[F : E] \leq \aleph_0$ means that every F-class contains only countably many E-classes.

We denote by $\Delta_X = \{(x, y) : x = y\}$ the equality relation on a set X, and we also let $I_X = X^2$. Note that if $E_y = E, y \in Y$, where E is an equivalence relation on a set X, then $\bigoplus_y E_y = E \times \Delta_Y$.

If $E_i, i \in I$, are equivalence relations on X, we denote by $\bigwedge_i E_i = \bigcap_i E_i$ the largest (under inclusion) equivalence relation contained in all E_i and by $\bigvee_i E_i$ the smallest (under inclusion) equivalence relation containing each E_i. We call $\bigwedge_i E_i$ the **meet** and $\bigvee_i E_i$ the **join** of (E_i).

If E is an equivalence relation on X, a set $S \subseteq X$ is a **complete section** of E if S intersects every E-class. Moreover, if S intersects every E-class in exactly one point, then S is a **transversal** of E.

Consider now an action $a : G \times X \to X$ of a group G on a set X. We often write $g \cdot x = a(g, x)$, if there is no danger of confusion. Let $G \cdot x = \{g \cdot x : g \in G\}$ be the **orbit** of $x \in X$. The action a induces an equivalence relation E_a on X whose classes are the orbits, i.e., $xE_a y \iff \exists g(g \cdot x = y)$. When a is understood, sometimes the equivalence relation E_a is also denoted by E_G^X. The action a is **free** if $g \cdot x \neq x$ for every $x \in X, g \in G, g \neq 1_G$.

1.2 Morphisms

Let E, F be equivalence relations on spaces X, Y, resp. A map $f : X \to Y$ is a **homomorphism** from E to F if $xEy \implies f(x)Ff(y)$. In this case we write $f : (X, E) \to (Y, F)$ or just $f : E \to F$, if there is no danger of confusion. A homomorphism f is a **reduction** if moreover $xEy \iff f(x)Ff(y)$. We denote this by $f : (X, E) \leq (Y, F)$ or just $f : E \leq F$. Note that a homomorphism as above induces a map from X/E to Y/F, which is an injection if f is a reduction. In other words, a homomorphism is a lifting of a map from X/E to Y/F, and a reduction is a lifting of an injection of X/E into Y/F. An **embedding** is an injective reduction. This is denoted by $f : (X, E) \sqsubseteq (Y, F)$ or just $f : E \sqsubseteq F$. An **invariant embedding** is an injective reduction whose range is an F-invariant subset of Y. This is denoted by $f : (X, E) \sqsubseteq^i (Y, F)$ or just $f : E \sqsubseteq^i F$. Finally, an **isomorphism** is a surjective embedding. This is denoted by $f : (X, E) \cong (Y, F)$ or just $f : E \cong F$.

If a, b are actions of a group G on spaces X, Y, resp., a **homomorphism** from a to b is a map $f : X \to Y$ such that $f(g \cdot x) = g \cdot f(x), \forall g \in G, x \in X$. If f is injective, we call it an **embedding** of a to b.

1.3 The Borel Category

We are interested here in studying (classes of) Borel equivalence relations on **standard Borel spaces** (i.e., Polish spaces with the associated Borel structure). If X is a standard Borel space space and E is an equivalence relation on X, then E is Borel if E is a Borel subset of X^2.

Given a class of functions Φ between standard Borel spaces, we can restrict the above notions of morphism to functions in Φ, in which case we use the subscript Φ in the above notation (e.g., $f\colon E \to_\Phi F$, $f\colon E \leq_\Phi F$, etc.). In particular if Φ is the class of Borel functions, we write $f\colon E \to_B F, f\colon E \leq_B F, f\colon E \sqsubseteq_B F, f\colon E \sqsubseteq^i_B F, f\colon E \cong_B F$ to denote that f is a Borel morphism of the appropriate type. Similarly when we consider the underlying topology, we use the subscript c in the case where Φ is the class of continuous functions between Polish spaces and write $f\colon E \to_c F, f\colon E \leq_c F, f\colon E \sqsubseteq_c F, f\colon E \sqsubseteq^i_c F, f\colon E \cong_c F$.

We say that E is **Borel reducible** to F if there is a Borel reduction from E to F. In this case we write $E \leq_B F$. If $E \leq_B F$ and $F \leq_B E$, then E, F are **Borel bireducible**, in symbols $E \sim_B F$. Finally we let $E <_B F$ if $E \leq_B F$ but $F \not\leq_B E$. Similarly we define the notions of E being **Borel embeddable** to F and E being **Borel invariantly embeddable** to F, for which we use the notations $E \sqsubseteq_B F$ and $E \sqsubseteq^i_B F$, respectively. Also we use $E \simeq_B F, E \simeq^i_B F$ for the corresponding notions of being **Borel biembeddable** and **Borel invariantly biembeddable** and $E \sqsubset_B F$ and $E \sqsubset^i_B F$ for the corresponding strict notions. More generally, if Φ is as above, we analogously define $E \leq_\Phi F, E \sqsubseteq_\Phi F$, etc.

Finally E, F are **Borel isomorphic**, in symbols $E \cong_B F$, if there is a Borel isomorphism from E to F. Note that by the usual (Borel) Schröder–Bernstein argument, E, F are Borel isomorphic if and only if they are Borel invariantly biembeddable, i.e., $\simeq^i_B = \cong_B$.

2

Countable Borel Equivalence Relations

Definition 2.1 An equivalence relation E is **countable** if every E-class is countable. It is **finite** if every E-class is finite.

2.1 Some Examples

We discuss first some examples of countable Borel equivalence relations.

Examples 2.2

(1) Let $X = 2^{\mathbb{N}}$. Then the **eventual equality** relation

$$xE_0y \iff \exists m\forall n \geq m(x_n = y_n)$$

and the **tail** equivalence relation

$$xE_ty \iff \exists m\exists k\forall n(x_{m+n} = y_{k+n})$$

are countable Borel.

More generally, let $(S, \times, 1)$ be a monoid. We usually write st for $s \times t$. An action a of S on a set X is a map $a\colon S \times X \to X$ such that, letting as usual $s \cdot x = a(s,x)$, we have $1 \cdot x = x, s \cdot (t \cdot x) = st \cdot x$. If now S is abelian, this action gives rise to two equivalence relations $E_{0,a}, E_{t,a}$ on X, defined by $xE_{0,a}y \iff \exists s(s \cdot x = s \cdot y), xE_{t,a}y \iff \exists s\exists t(s \cdot x = t \cdot y)$. If we take $S = (\mathbb{N}, +, 0)$, $X = 2^{\mathbb{N}}, 1 \cdot (x_n) = (x_{n+1})$ (the shift map), we obtain E_0, E_t. If S is countable (discrete) and abelian, and a is a Borel action such that for each $s \in S$ the map $x \mapsto s \cdot x$ is countable-to-1, then $E_{0,a}, E_{t,a}$ are countable Borel.

(2) Take again $X = 2^{\mathbb{N}}$ and consider $\equiv_T$ and $\equiv_A$, the **Turing** and **arithmetical equivalence** relations, resp. These are countable Borel.

(3) Let now $X = \mathbb{R}$. Then the **Vitali equivalence relation** defined by $xE_vy \iff x - y \in \mathbb{Q}$ is countable Borel.

(4) Let $X = \mathbb{R}^+$. The **commensurability relation** is given by $xE_c y \iff \frac{x}{y} \in \mathbb{Q}$. This is countable Borel (and one of the earliest equivalence relations in the history of mathematics).

(5) Let $k \geq 2$, and let X be the space of subshifts of $k^{\mathbb{Z}}$, where a **subshift** is a nonempty closed subset of $k^{\mathbb{Z}}$ invariant under the shift map $S(x)_i = x_{i-1}$. This is a compact subspace of the hyperspace of all compact subsets of $k^{\mathbb{Z}}$, thus compact and metrizable. Let E be the equivalence relation of isomorphism of subshifts, where two subshifts are isomorphic if there is a homeomorphism between the closed sets that commutes with the shift. Then E is a countable Borel equivalence relation, see [Cl2].

(6) Now let a be a Borel action of a countable (discrete) group G on a standard Borel space X. Then E_a is a countable Borel equivalence relation.

2.2 The Feldman–Moore Theorem

It turns out that (6) in the list of Examples 2.2 includes all countable Borel equivalence relations.

Theorem 2.3 ([FM]) *If E is a countable Borel equivalence relation on a standard Borel space X, then there are a countable group G and a Borel action a of G on X such that $E = E_a$.*

This is an immediate consequence of the following result that can be proved using the classical Lusin–Novikov theorem in descriptive set theory, see [Ke6, 18.10].

Theorem 2.4 ([FM]) *If E is a countable Borel equivalence relation on a standard Borel space X, then there is a sequence of Borel involutions (T_n) on X such that $xEy \iff \exists n(T_n(x) = y)$.*

Remark 2.5 In Theorem 2.4 one can also find (T_n) as in that theorem such that moreover for any $x \neq y, xEy$, there is a unique n such that $T_n(x) = y$. This is equivalent to saying that the Borel graph $E \setminus \Delta_X$ has a countable Borel edge chromatic number and follows from the general result [KST, 4.10] (see also [Ke11, 3.7]).

Although Theorem 2.3 always guarantees the existence of a Borel action of a countable group that generates a given countable Borel equivalence relation, it is not always clear how to find a "natural" such action that generates a specific equivalence relation of interest. Considering the items listed in Examples 2.2, E_0 is generated by an action as follows. Identify $2^{\mathbb{N}}$ with the compact product group $(\mathbb{Z}/2\mathbb{Z})^{\mathbb{N}}$ and consider the translation action of its countable dense subgroup $(\mathbb{Z}/2\mathbb{Z})^{<\mathbb{N}}$. The induced equivalence relation is clearly E_0. It follows from a general result of [GPS, 3.9] that E_0 is generated by a continuous $\mathbb{Z}$-action, i.e., it is generated by a single homeomorphism. A more direct construction is given in [Cl1]. (The action of $\mathbb{Z}$ on $2^{\mathbb{N}}$ induced by the **odometer** map, i.e., addition of 1 modulo 2 with right carry, generates an equivalence relation in which one class consists of the eventually constant sequences and the others coincide with the E_0-classes.) Joshua Frisch pointed out that if $T: 2^{\mathbb{N}} \to 2^{\mathbb{N}}$ is the homeomorphism given by $01x \to 10x, 00x \to 0x, 1x \to 11x$, for $x \in 2^{\mathbb{N}}$, and U is the homeomorphism given by $0x \to 1x, 1x \to 0x$, for $x \in 2^{\mathbb{N}}$, then the group $\langle T, U\rangle$ generates E_t. As opposed to E_0, E_t cannot be generated by a single homeomorphism, as such would have an invariant probability Borel measure (by the amenability of $\mathbb{Z}$), but it can be shown that any Borel action of a countable group that generates E_t cannot have such an invariant measure (see the paragraph preceding Corollary 4.7). We will see in Example 7.7(1) that E_t can be generated by a single Borel automorphism.

The Vitali equivalence relation and the commensurability relation are clearly generated by actions of $(\mathbb{Q}, +)$ and $(\mathbb{Q}^+, \cdot)$, resp. It is not clear how to explicitly find actions that generate $E_{0,a}, E_{t,a}, \equiv_T, \equiv_A$ and isomorphism of subshifts.

Definition 2.6 A finite equivalence relation E is of **type** n if every E-class has cardinality $\leq n$.

The following is an immediate corollary of Theorem 2.4:

Corollary 2.7 *If E is a countable Borel equivalence relation, then there is a sequence (E_n) of Borel equivalence relations of type 2 such that $E = \bigvee_n E_n = \bigcup_n E_n$.*

It is natural to ask whether in Corollary 2.7 one can find finitely many *finite* Borel equivalence relations $(E_n)_{n<N}$ with $E = \bigvee_{n<N} E_n$. This is, however, ruled out by the theory of cost, see [Ga1]. The following is shown in [JKL]:

Theorem 2.8 ([JKL, 1.21]) *For every countable Borel equivalence relation E, there is a countable Borel equivalence relation F such that $E \sim_B F$ and F is of the form $F = G \vee H$, where G, H are Borel equivalence relations of types 2, 3, resp. Moreover, this fails if we require that such F, G, H can be found, where G, H are of type 2. However one can write such an F as $F = G \vee H \vee K$, with G, H, K of type 2.*

In fact it is shown in [JKL, 1.21] that the equivalence relations of the form $G \vee H$, with G, H of type 2, are exactly the hyperfinite ones, see Chapter 7.

Here a countable Borel equivalence relation E is **hyperfinite** if it can be written as $E = \bigcup_n E_n$, where (E_n) is an increasing sequence ($E_n \subseteq E_{n+1}$, for each n) of finite Borel equivalence relations.

2.3 Induced Actions

There is a very useful construction, called the **inducing construction**, due to Mackey, that allows for a pair of groups, $G \leq H$, to extend an action of a group G to an action of H.

Theorem 2.9 (see [BK, 2.3.5]) *Let H be a Polish group and $G \leq H$ a closed subgroup.*

(a) (Mackey) *Let a be a Borel action of G on a standard Borel space X. Then there is a standard Borel space Y such that $X \subseteq Y$ and X is a Borel subset of Y, and a Borel action b of H on Y, with the following properties:*

(i) *For $x \in X$ and $g \in G$, $a(g,x) = b(g,x)$.*
(ii) *Every orbit of H on Y contains exactly one orbit of G on X.*
(iii) *$E_a \sqsubseteq_B E_b$ and $E_b \leq_B E_a$; therefore $E_a \sim_B E_b$.*
(iv) *If a is a free action, so is b.*

(b) (Hjorth) *If a as above is a continuous action of G on a Polish space X, then Y can be taken to be also Polish and the action b continuous, and moreover X a closed subspace of Y.*

The action b is called the **induced action** of a and is denoted by $\mathrm{IND}_G^H(a)$.

2.4 Closure Properties

We record in the following some simple closure properties of the class of countable Borel equivalence relations.

Proposition 2.10

(i) *If E is a countable Borel equivalence relation on X and $A \subseteq X$ is Borel, then $E \upharpoonright A$ is also countable Borel.*
(ii) *If F is a countable Borel equivalence relation and $E \sqsubseteq_B F$, then E is also countable Borel.*
(iii) *If $E \subseteq F$ are Borel equivalence relations and F is countable, so is E. If E is countable and $[F : E] \leq \aleph_0$, then F is countable.*

(iv) *If E, F are countable Borel equivalence relations, then so is $E \times F$.*

(v) *If each $E_n, n \in \mathbb{N}$, is a countable Borel equivalence relation, then so is $\bigoplus_n E_n$. More generally, If Y is a standard Borel space and $(E_y)_{y \in Y}$ is a family of countable Borel equivalence relations on a standard Borel space X such that $\{(x, u, y) \in X^2 \times Y : (x, u) \in E_y\}$ is Borel, then $\bigoplus_y E_y$ is countable Borel.*

(vi) *If each $E_n, n \in \mathbb{N}$, is a countable Borel equivalence relation, then so is $\bigvee_n E_n$.*

For a standard Borel space and a Borel map $T : X \to X$, let E_T be the smallest equivalence relation containing the graph of T. If T is countable-to-1, then E_T is countable Borel (being equal to $E_{t,a}$, where a is the Borel action of $(\mathbb{N}, +, 0)$ generated by T, i.e., $1 \cdot x = T(x)$, see Example 2.2(1). More generally, let T_n be a sequence of countable-to-1 Borel maps from X to X, and let $E_{(T_n)}$ be the smallest equivalence relation containing the graphs of all T_n. Then $E_{(T_n)} = \bigvee_n E_{T_n}$ is countable Borel.

2.5 Complete Sections and Vanishing Sequences of Markers

If E is a countable Borel equivalence relation on a standard Borel space X and A is a Borel complete section for A, then we view $A \cap [x]_E$ as putting a set of markers on the E-class of x in a uniform Borel way, so sometimes we call such an A a **marker set**. Finding appropriate marker sets plays an important role in the study of countable Borel equivalence relations.

The simplest situation is when a Borel transversal can be found.

Definition 2.11 A Borel equivalence relation E on a standard Borel space X is called **smooth** if there is a Borel function $f : X \to Y$, Y a standard Borel space, such that $xEy \iff f(x) = f(y)$, i.e., $E \leq_B \Delta_Y$.

For example, any finite Borel equivalence relation is smooth. We now have the following basic fact:

Proposition 2.12 *The following are equivalent for a countable Borel equivalence relation E:*

(i) *E is smooth.*

(ii) *E admits a Borel transversal.*

(iii) *The space X/E with the* **quotient Borel structure** *Σ_E (i.e., $A \subseteq X/E \in \Sigma_E \iff \tilde{A} = \bigcup A \subseteq X$ is Borel) is standard.*

For example, the equivalence relations in Examples 2.2(1–5), except possibly for $E_{0,a}, E_{t,a}$, for some a, are not smooth.

We will next discuss a very different characterization of smoothness.

A **mean** on a countable set S is a positive linear functional $\varphi : \ell^\infty(S) \to \mathbb{C}$, which assigns the value 1 to the constant 1 function. Let E be a countable Borel equivalence relation on a standard Borel space X. An **assignment of means** is a map that associates to each equivalence class $[x]_E$ a mean $\varphi_{[x]_E}$ on $[x]_E$. Finally, an assignment of means $[x]_E \mapsto \varphi_{[x]_E}$ is Borel (in the weak sense) if for each bounded Borel map $f \colon E \to \mathbb{C}$, the function $x \mapsto \varphi_{[x]_E}(f_x)$ is Borel, where $f_x(y) = f(x, y)$.

Theorem 2.13 ([KM2]) *Let E be a countable Borel equivalence relation. Then the following are equivalent:*

(i) *E is smooth.*
(ii) *E admits a Borel assignment of means.*

Remark 2.14 An analog of Theorem 2.13 in the Baire category context is also proved in [KM2]. We will see in Section 8.4 that in the measure-theoretic context the situation is quite different, since smoothness is replaced in this case by hyperfiniteness.

The next, very useful, result guarantees the existence of appropriate markers even in the nonsmooth situation. An equivalence relation E is called **aperiodic** if every E-class is infinite.

Theorem 2.15 (The marker lemma, [SlSt]) *Let E be an aperiodic countable Borel equivalence relation on a standard Borel space X. Then E admits a vanishing sequence of Borel markers, i.e., there is a sequence of complete Borel sections (A_n), with $A_0 \supseteq A_1 \supseteq A_2 \dots$ and $\bigcap_n A_n = \emptyset$.*

From this we also have the following:

Corollary 2.16 *Let E be an aperiodic countable Borel equivalence relation on a standard Borel space X. Then E admits a pairwise disjoint sequence of Borel markers, i.e., there is a sequence of complete Borel sections (B_n), with $B_n \cap B_m = \emptyset$, if $m \neq n$.*

For a proof, see, e.g., [CM1, 1.2.6]. A generalization of Theorem 2.15 to transitive Borel binary relations with countably infinite vertical sections can be found in [Mi3].

It is also clear that for any finite set of aperiodic countable Borel equivalence relations there is a common vanishing sequence of Borel markers. Concerning common vanishing sequences of Borel markers for infinite sets of aperiodic countable Borel equivalence relations, we have the following results:

Proposition 2.17

(i) (Miller) *There is a sequence of aperiodic countable Borel equivalence relations* (F_n) *in a Polish space* X*, such that if* $A \subseteq X$ *is a Borel complete section for all* F_n*, then* A *is comeager. Thus for any sequence of Borel sets* (A_m) *such that each* A_m *is a complete section for all* F_n*,* $\bigcap_m A_m$ *is comeager. In particular the sequence* (F_n) *has no common vanishing sequence of Borel markers.*

(ii) (Marks) *Let* (F_n) *be a sequence of aperiodic countable Borel equivalence relations on a standard Borel space* X*, and let* μ *be a probability Borel measure on* X*. Then there is a decreasing sequence of Borel sets* (A_m) *such that each* A_m *is a complete section for every* F_n *and* $\mu(\bigcap_n A_n) = 0$.

Proof (i) Let F_n be the subequivalence relation of E_0 defined by $xF_ny \iff xE_0y \ \& \ (x_i)_{i<n} = (y_i)_{i<n}$. Then if a Borel set $A \subseteq X$ is a complete section for F_n, A is nonmeager in every basic neighborhood (nbhd) $N_s = \{x \in 2^{\mathbb{N}} : (x_i)_{i<n} = s\}$, where $s \in 2^n$. Thus if a Borel set A is a complete section for all F_n, A must be comeager.

(ii) Let $(A_{n,m})$ be a vanishing sequence of Borel markers for E_n such that $\mu(A_{n,m}) \leq \frac{1}{2^{n+m}}$. Put $A_m = \bigcup_n A_{n,m}$. □

An important "dual" question (especially because of its connection to Borel combinatorics, see [M1]) is whether two countable Borel equivalence relations can have disjoint complete sections. Here we have the following results:

Theorem 2.18 ([M1, Section 4])

(i) *There are aperiodic countable Borel equivalence relations* E, F *on a standard Borel space* X *such that there is no Borel set* $A \subseteq X$ *with* A *a complete section for* E *and* $X \setminus A$ *a complete section for* F.

(ii) *For any two countable Borel equivalence relations* E, F *on a standard Borel space* X *such that all* E*-classes have cardinality at least* 3 *and all* F*-classes have cardinality at least* 2*, and for every probability Borel measure* μ *on* X*, there is Borel* $A \subseteq X$ *such that* A *meets* μ*-a.e.* E *class (i.e.,* $\mu([A]_E) = 1$*) and* $X \setminus A$ *meets* μ*-a.e.* F *class.*

(iii) *For any two countable Borel equivalence relations* E, F *on a Polish space* X *such that all* E*-classes have cardinality at least* 3 *and all* F*-classes have cardinality at least* 2*, there is Borel* $A \subseteq X$ *such that* A *meets comeager many* E*-classes (i.e.,* $[A]_E$ *is comeager) and* $X \setminus A$ *meets comeager many* F*-classes.*

Finally, for certain countable Borel equivalence relations, and especially those generated by shift actions of groups (see Section 5.3), there are several interesting results concerning the topological structure of vanishing sequences of markers and the local structure of complete sections; see [GJS1], [GJKS], [M3], and [CMa].

2.6 Maximal Finite Partial Subequivalence Relations

Another useful tool in studying countable Borel equivalence relations is the existence of appropriate finite partial subequivalence relations. Here by a **partial subequivalence relation** of an equivalence relation E on a space X, we mean an equivalence relation F on a subset $A \subseteq X$ such that $F \subseteq E$. A *finite* partial subequivalence relation is abbreviated as **fsr**.

Let now X be a standard Borel space, and denote by $[X]^{<\infty}$ the standard Borel space of finite subsets of X. If E is an equivalence relation on X, we denote by $[E]^{<\infty}$ the subset of $[X]^{<\infty}$ consisting of all finite sets that are contained in a single E-class. If E is Borel, so is $[E]^{<\infty}$. For each set $\Phi \subseteq [E]^{<\infty}$, an fsr F of E defined on the set $A \subseteq X$ is **Φ-maximal** if every F-class is in Φ and every finite set S disjoint from A is not in Φ. We now have the following that is proved using a result from Borel combinatorics.

Theorem 2.19 ([KM1, 7.3]) *If E is a countable Borel equivalence relation and $\Phi \subseteq [E]^{\infty}$ is Borel, then there is a Borel Φ-maximal fsr of E.*

The following is a typical application of this result.

Corollary 2.20 *Let (M_n) be a sequence of positive integers ≥ 2. Then for each aperiodic countable Borel equivalence relation E, there is an increasing sequence of finite Borel subequivalence relations (E_n) of E such that each E_n-class has exactly $M_0 M_1 \cdots M_n$ elements.*

Proof It is shown in [KM1, 7.4] (using Theorem 2.19) that given a positive integer M, every aperiodic countable Borel equivalence contains a finite subequivalence relation all of whose classes have cardinality M. One can then define E_n inductively as follows: Given E_n, let X_n be a Borel transversal for E_n. Apply this fact to $E \upharpoonright X_n$ to find a finite subequivalence relation $F_n \subseteq E \upharpoonright X_n$, each of whose classes has cardinality M_{n+1}, and then take $E_{n+1} = E_n \vee F_n$. □

If in Corollary 2.20 we let $F = \bigcup_n E_n$, then F is an aperiodic hyperfinite Borel subequivalence relation of E.

2.7 Compressibility

Recall that a set C is called **Dedekind infinite** if there is an injection $f\colon C \to C$ such that $f(C) \subsetneq C$, i.e., C can be compressed into a proper subset of itself. The following is an analog of this concept in the context of countable Borel equivalence relations.

Definition 2.21 Let E be a countable Borel equivalence relation on a standard Borel space X. We say that E is **compressible** if there is an injective Borel map f: $X \to X$ such that for each E-class C, $f(C) \subsetneq C$. Such a map is called a (Borel) **compression**. A Borel set $A \subseteq X$ is compressible if $E \restriction A$ is compressible.

For any countable Borel equivalence relation E on a standard Borel space X and Borel sets $A, B \subseteq X$, we let

$$A \sim_E B \iff \exists f\colon A \to B(f \text{ is a Borel bijection and } \forall x \in A(f(x)Ex)).$$

In particular, $A \sim_E B \implies E \restriction A \cong_B E \restriction B$. We also put

$$A \preceq_E B \iff \exists \text{ Borel } \ C \subseteq B(A \sim_E C)$$

and

$$A \prec_E B \iff \exists \text{Borel } \ C \subseteq B(A \sim_E C, B \setminus C \text{ a complete section of } E|[B]_E).$$

The standard Borel Schröder–Bernstein argument shows that

$$A \sim_E B \iff A \preceq_E B \;\&\; B \preceq_E A.$$

Note also that a Borel set A is compressible if and only if $A \prec_E A$.

We also have the following, which is part of the proof of Theorem 4.6 in Chapter 4. See also [BK, 4.5.1]:

Proposition 2.22 *Let E be a countable Borel equivalence relation on a standard Borel space X. Let A, B be two complete Borel sections for E. Then there is a partition $X = P \sqcup Q$ into E-invariant Borel sets such that $A \cap P \prec_E B \cap P$ and $B \cap Q \preceq_E A \cap Q$.*

A Borel set $A \subseteq X$ is called ***E*-paradoxical** if there are disjoint Borel subsets $B, C \subseteq A$ such that $A \sim_E B, A \sim_E C$.

The following result, for which we refer to [DJK, Section 2] and references therein to [CN1], [CN2], [N1], and [N2], gives a number of equivalent formulations of compressibility.

Proposition 2.23 *Let E be a countable Borel equivalence relation on a standard Borel space X. Then the following are equivalent:*

(i) *E is compressible.*
(ii) *There is a sequence of pairwise disjoint complete Borel sections (A_n) of E such that $A_i \sim_E A_j$ for each i, j.*
(iii) *There is an infinite partition $X = A_0 \sqcup A_1 \sqcup \cdots$ into complete Borel sections such that $A_i \sim_E A_j$ for each i, j.*
(iv) *The space X is E-paradoxical.*
(v) *$E \cong_B E \times I_{\mathbb{N}}$.*
(vi) *There is a smooth aperiodic Borel subequivalence relation $F \subseteq E$.*

We call $E \times I_{\mathbb{N}}$ the **amplification** of E. Thus the compressible equivalence relations are those that are Borel isomorphic to their amplifications. Note that for any countable Borel equivalence relation E, $E \sim_B E \times I_{\mathbb{N}}$.

Remark 2.24 In contrast to Proposition 2.23(iii), from [KM1, 7.4] (see also Corollary 2.20) it follows that for *any* aperiodic countable Borel equivalence relation E on a standard Borel space X and for any $n \geq 1$, there is a finite partition $X = A_0 \sqcup A_1 \sqcup \cdots \sqcup A_{n-1}$ into complete Borel sections such that $A_i \sim_E A_j$ for each $i, j < n$. See also [S15, 1.8.5].

Another characterization of compressibility is the following, where for a countable Borel equivalence relation E on a standard Borel space X, a Borel set $A \subseteq X$ is called ***E*-syndetic** if for some $n > 0$ there are Borel sets $A_i, i < n$, such that $A \sim_E A_i, \forall i < n$, and $X = \bigcup_{i<n} A_i$.

Proposition 2.25 ([S12, Proposition 10.2]) *Let E be an aperiodic countable Borel equivalence relation on a standard Borel space X. Then the following are equivalent:*

(i) *E is compressible.*
(ii) *For any two Borel syndetic sets $A, B \subseteq X$, $A \sim_E B$.*

In Chapter 4 we will also see Nadkarni's characterization of compressibility in terms of lack of invariant measures.

It is easy to see that $E_t, E_v, E_c, \equiv_T, \equiv_A$ are compressible and so is the eventual equality relation $E_0(\mathbb{N})$ on $\mathbb{N}^{\mathbb{N}}$ (i.e., $xE_0(\mathbb{N})y \iff \exists m \forall n \geq m(x_n = y_n)$). However, E_0 is not compressible (see Section 4.3).

The following is also a basic fact concerning compressible sets, see [N1, 5.7] or [DJK, 2.2].

Proposition 2.26 *Let E be a countable Borel equivalence relation on a standard Borel space X. If a Borel set $A \subseteq X$ is compressible, then we have that $A \sim_E [A]_E$. Thus $[A]_E$ is also compressible.*

The next result deals with embeddability for compressible relations.

Proposition 2.27 *Let E, F be countable Borel equivalence relations.*

(i) *If E is compressible, then $E \sqsubseteq_B F \iff E \sqsubseteq^i_B F$.*
(ii) *If both E, F are compressible, then $E \leq_B F \iff E \sqsubseteq^i_B F$. In particular, if both E, F are compressible, then $E \sim_B F \iff E \simeq_B F \iff E \cong_B F$.*

Part (i) of Proposition 2.27 follows from Proposition 2.26. For part (ii), see [CK, 5.23].

The following gives another connection between reducibility and embeddability.

Proposition 2.28 *Let E, F be countable Borel equivalence relations on standard Borel spaces X, Y, resp. Then*

$$E \leq_B F \iff E \sqsubseteq_B F \times I_{\mathbb{N}}.$$

Proof If $E \sqsubseteq_B F \times I_{\mathbb{N}}$, then clearly $E \leq_B F$, since $F \times I_{\mathbb{N}} \leq_B F$. If now $E \leq_B F$, let $f: X \to Y$ be a Borel reduction of E to F. Then $f(X) = A$ is a Borel subset of Y, and there is sequence of Borel maps $g_n: A \to X$ such that $(g_n(y))$ enumerates $f^{-1}(y)$ for each $y \in A$. Let $g: X \to Y \times \mathbb{N}$ be defined by $g(x) = (f(x), i)$, where i is least such that $g_i(f(x)) = x$. This witnesses that $E \sqsubseteq_B F \times I_{\mathbb{N}}$. □

Finally it turns out that generically *every* aperiodic countable Borel equivalence relation is compressible.

Theorem 2.29 ([KM1, 13.3]) *Let E be an aperiodic countable Borel equivalence relation on a Polish space X. Then there is an invariant comeager Borel set $C \subseteq X$ such that $E \upharpoonright C$ is compressible.*

For a stronger result involving graphings of equivalence relations, see [CKM, Section 4]. Also for related results about semigroup actions, see [Mi7].

Remark 2.30 We say that an infinite countable group G is **dynamically compressible** if every aperiodic E generated by a Borel action of G can be Borel reduced to a compressible aperiodic F induced by a Borel action of G. It is shown in [FKSV] that every infinite countable amenable group is dynamically compressible, and the same is true for any countable group that contains a nonabelian free group. However, there are infinite countable groups that fail to satisfy these two conditions but are still dynamically compressible (see again [FKSV]). It is not known if *every* infinite countable group is dynamically compressible.

Remark 2.31 For a connection between compressibility and cardinal algebras (discussed in Chapter 12), see Remark 12.5.

2.8 Borel Cardinalities and a Schröder–Bernstein-Type Theorem

If E, F are countable Borel equivalence relations on standard Borel spaces X, Y, resp., then $E \leq_B F$ means that there is an injective map

$$f\colon X/E \to Y/F$$

that has a Borel lifting. We can interpret this as meaning that the **Borel cardinality** of X/E is at most that of Y/F, in symbols

$$|X/E|_B \leq |Y/F|_B.$$

We also let

$$|X/E|_B = |Y/F|_B \iff |X/E|_B \leq |Y/F|_B \ \&\ |Y/F|_B \leq |X/E|_B,$$

so that $|X/E|_B = |Y/F|_B \iff E \sim_B F$, and

$$|X/E|_B < |Y/F|_B \iff |X/E|_B \leq |Y/F|_B \ \&\ |Y/F|_B \nleq |X/E|_B,$$

so that $|X/E|_B < |Y/F|_B \iff E <_B F$.

The next result provides analogs of the classical Schröder–Bernstein theorem that in particular show that $X/E, Y/F$ have the same Borel cardinality (i.e., $|X/E|_B = |Y/F|_B$) if and only if there is a bijection between X/E and Y/F with Borel lifting.

Theorem 2.32 *Let E, F be countable Borel equivalence relations on standard Borel spaces X, Y, resp. Then the following are equivalent:*

(i) $E \sim_B F$.
(ii) *There are Borel sets $A \subseteq X, B \subseteq Y$ that are complete sections of E, F, resp., such that $E \restriction A \cong_B F \restriction B$.*
(iii) $E \times I_{\mathbb{N}} \cong_B F \times I_{\mathbb{N}}$.
(iv) *There is a bijection $f\colon X/E \to Y/F$ with Borel lifting (in which case f^{-1} has also a Borel lifting).*
(v) *There are decompositions $X = X_1 \sqcup X_2, Y = Y_1 \sqcup Y_2$ into invariant Borel sets and Borel complete sections $A_2 \subseteq X_2, B_1 \subseteq Y_1$ of $E \restriction X_2, F \restriction Y_1$, resp., such that $E \restriction X_1 \cong_B F \restriction B_1$ and $F \restriction Y_2 \cong_B E \restriction A_2$.*

For the proof of the equivalence of parts (i)–(iv) see [DJK, 2.6], and for (v) see [Mi1]. Alternatively, as pointed out by Ronnie Chen [oral communication], one can see that (iv) implies (v) as follows: Let Z be the disjoint union of the spaces X, Y and define the Borel equivalence relation R on Z by gluing together any E-class C with the F-class $f(C)$, where f is as in (iv). Then X, Y are complete Borel sections for R, and an application of Proposition 2.22 gives (v).

It was asked in [DJK, page 201] whether Borel bireducibility, for aperiodic countable Borel equivalence relations, is also equivalent to Borel biembeddability. This is equivalent to asking whether $E \times I_{\mathbb{N}} \sqsubseteq_B E$ holds for all aperiodic countable Borel equivalence relations E. This was disproved by Simon Thomas in [T2], using methods of ergodic theory.

Theorem 2.33 ([T2]) *There is an aperiodic countable Borel equivalence relation E such that it is not the case that $E \times I_2 \sqsubseteq_B E$.*

Other proofs of this theorem can be found in [HK4, 3.9] and [CM1, Theorem H].

For a countable Borel equivalence relation E on a standard Borel space X, recall that the quotient Borel structure on X/E, Σ_E, is the σ-algebra on X/E defined by: $A \in \Sigma_E \iff \tilde{A} = \bigcup A (\subseteq X)$ is Borel. We say that two countable Borel equivalence relations E, F on X, Y, resp., are **quotient Borel isomorphic** if there is a bijection of X/E to F/Y that takes Σ_E to Σ_F. Denote this by $E \cong^q_B F$. It is easy to see that $E \sim_B F \implies E \cong^q_B F$.

Problem 2.34 Is it true that $E \sim_B F \iff E \cong^q_B F$?

Finally we note the following result about liftings.

Proposition 2.35 ([Mi5, Proposition 6.1]) *Let E, F be countable Borel equivalence relations on standard Borel spaces X, Y, resp., and assume that F is aperiodic. Let $f\colon X/E \to Y/F$ be a countable-to-1 function. Then if f has a Borel lifting, it has a finite-to-1 Borel lifting.*

2.9 Weak Borel Reductions

We now consider a weaker notion of Borel reduction.

Definition 2.36 Let E, F be countable Borel equivalence relations. A **weak Borel reduction** of E to F is a countable-to-1 Borel homomorphism f from E to F. We denote this by $f\colon E \leq^w_B F$. If such an f exists, we say that E is **weakly Borel reducible** to F, in symbols $E \leq^w_B F$.

Clearly when $E \subseteq F$, the identity map is a weak Borel reduction of E to F. Of course a Borel reduction is also a weak Borel reduction. It turns out that weak reducibility is just the combination of inclusion and reduction. More precisely we have the following result, attributed to Kechris and Miller in [T12].

Theorem 2.37 (see [T12, Section 4]) *Let E, F be countable Borel equivalence relations on uncountable standard Borel spaces X, Y, resp. Then the following are equivalent:*

(i) $E \leq_B^w F$.
(ii) *There is a countable Borel equivalence relation $E' \supseteq E$ with $E' \leq_B F$.*
(iii) *There is a Borel subequivalence relation $F' \subseteq F$ such that $E \leq_B F'$.*

The notion of weak reduction is in general weaker than reduction. In fact we have the following stronger result, proved by using methods of ergodic theory.

Theorem 2.38 ([A6]) *There exist countable Borel equivalence relations E, F such that $E \subseteq F$ but $E \nleq_B F$ and $F \nleq_B E$.*

Other proofs of this result were given in [HK4, 3.8], [T12, Section 5], [H12] (see also [Mi11, 6.1]), and [CM1, Theorem G].

2.10 The Full Group and the Automorphism Group

To each countable Borel equivalence relation E one can assign a group of Borel automorphisms of the underlying space that actually determines it up to Borel isomorphism, at least in the aperiodic case.

Definition 2.39 Let E be a countable Borel equivalence relation on a standard Borel space X. The **full group** of E, in symbols $[E]_B$, is the group of all Borel automorphisms T of X such that $T(x)Ex, \forall x$.

Sometimes the full group is called the **inner automorphism group** of E and denoted by $\mathrm{Inn}_B(E)$.

We then have the following "Borel" analog of the classical theorem of Dye on (measure-theoretic) full groups of ergodic, measure-preserving countable Borel equivalence relations; see, e.g., [Ke10, 4.1].

Theorem 2.40 ([MR1]) *Let E, F be aperiodic countable Borel equivalence relations. Then the following are equivalent:*

(i) $E \cong_B F$.
(ii) *$[E]_B, [F]_B$ are isomorphic (as abstract groups).*

Moreover, for any (algebraic) isomorphism $\varphi\colon [E]_B \to [F]_B$, there is a Borel isomorphism $f\colon E \cong_B F$ such that for $T \in [E]$, we have $\varphi(T) = f \circ T \circ f^{-1}$.

In fact in [MR1] it is shown that Theorem 2.40 holds for more general equivalence relations than those that are countable Borel.

A further study of full groups can be found in [Me], [Mi5, Chapter 1], [Mi2], and [Mi16]. In particular one can characterize compressibility in terms of the algebraic properties of the full group.

Let G be a group. Then G has the **Bergman property** if for any increasing sequence $A_0 \subseteq A_1 \subseteq A_2 \subseteq \cdots \subseteq G$ with $\bigcup_n A_n = G$, there exist n, k such that $G = (A_n)^k$. It has the **strong Bergman property** if there is k such that for any increasing sequence $A_0 \subseteq A_1 \subseteq A_2 \subseteq \cdots \subseteq G$ with $\bigcup_n A_n = G$, there is n such that $G = (A_n)^k$.

We now have:

Theorem 2.41 ([Mi5, Theorem 8.18], [Mi16, Theorems 7,9]) *Let E be an aperiodic countable Borel equivalence relation. Then $[E]_B$ has the Bergman property, and the following are equivalent:*

(i) *E is compressible.*
(ii) *$[E]_B$ has the strong Bergman property.*

More recently Miller has obtained a very interesting first-order characterization of noncompressibility. In what follows, for a group G and $g \in G$, we denote by $\mathrm{Cl}(g)$ the conjugacy class of g.

Theorem 2.42 ([Mi18, Theorem 2]) *Let E be an aperiodic countable Borel equivalence relation, and let $n \geq 5$. Then the following are equivalent:*

(i) *E is not compressible.*
(ii) *There is $T \in [E]_B$ such that n is least such that $[E]_B = \mathrm{Cl}(T)^n$.*

Moreover, in [Mi18, Theorem 13] it is shown that the condition $n \geq 5$ is optimal.

Finally Rosendal has shown that the full group of any aperiodic countable Borel equivalence relation on an uncountable standard Borel space (in fact any countable Borel equivalence relation with uncountably many classes of cardinality at least 3) admits no second countable Hausdorff topology in which it becomes a topological group. The proof is similar to that of [Ro, Theorem 1].

The normalizer of $[E]_B$ in the group of all Borel automorphisms of X is denoted by $N_B[E]$. It consists exactly of all Borel automorphisms of E, i.e., all Borel automorphisms T of X such that $xEy \iff T(x)ET(y)$, for all $x, y \in X$. For this reason, $N_B[E]$ is sometimes called the **automorphism group** of E and denoted by $\mathrm{Aut}_B(E)$. Moreover, by Theorem 2.40, for aperiodic E, the group $N_B[E]$ can be identified with the group of automorphisms of the group $[E]_B$.

The quotient group $N_B[E]/[E]_B$ is called the **outer automorphism group** of E, in symbols $\mathrm{Out}_B(E)$.

Consider now a Borel action a of a countable group G on X by automorphisms of E, i.e., for each $g \in G$, $x \mapsto g \cdot x$ is in $N_B[E]$. We abbreviate this by writing $a\colon G \curvearrowright_B (X, E)$. Then we can define the following equivalence relation on X, denoted by $E(a)$:

$$xE(a)y \iff \exists g \in G(g \cdot xEy).$$

Clearly $E(a) = E \vee E_a$. We call E_a the **expansion** of E by the action a. The following is due to J. Frisch and F. Shinko:

Proposition 2.43 *Every countable Borel equivalence relation is Borel bireducible to an expansion of a smooth countable Borel equivalence relation.*

Proof Let E be a countable Borel equivalence relation on X, and let a be a Borel action of a countable group G on X with $E = E_a$. Consider the action b of G on $X \times G$ given by: $g \cdot (x, h) = (g \cdot x, gh)$. Let also F on $X \times G$ be defined by : $(x, g)F(y, h) \iff x = y$, so that F is smooth and clearly G acts by automorphisms of F. Then note that $(x, g)F(b)(y, h) \iff xEy$, so that $E \sim_B F(b)$. □

If $E \subseteq F$ are countable Borel equivalence relation on X, we say that E is **normal** in F (or F is normal over E) if F is an expansion of E, i.e., there is $a\colon G \curvearrowright_B (X, E)$ with $F = E(a)$. Thus up to bireducibility, every countable Borel equivalence relation is normal over a smooth one.

2.11 Actions on Quotient Spaces

Let E, F be countable Borel equivalence relations on X, Y, resp. We call a map $f\colon X/E \to Y/F$ **Borel** if it has a Borel lifting $\tilde{f}\colon X \to Y$. We denote by $\mathrm{Sym}_B(X/E)$ the **Borel symmetric group** of X/E, i.e., the group of Borel permutations of X/E. Let G be a countable group. A **Borel action** of G on X/E, in symbols $G \curvearrowright_B X/E$, is an action of G on X/E by Borel permutations. For such an action, we let $E^{\vee G}$ be the countable Borel equivalence relation given by

$$xE^{\vee G}y \iff \exists g \in G(g \cdot [x]_E = [y]_E).$$

Note that an action $a\colon G \curvearrowright_B (X, E)$ gives canonically a Borel action of G on X/E given by $g \cdot [x]_E = [g \cdot x]_E$. Then it is easy to check that $E^{\vee G} = E(a)$.

(1) Given countable Borel equivalence relations $E \subseteq F$ on X, we say that F/E is **generated by a Borel action** if there is $G \curvearrowright_B X/E$ such that $F = E^{\vee G}$, i.e., if there is an analog of the Feldman–Moore theorem for X/F over X/E. In [dRM1, proof of Theorem 5] it is shown that there is a countable set of obstructions for being generated by a Borel action. Namely, there is a sequence of pairs $E_n \subseteq F_n$ of countable Borel equivalence relations on $2^{\mathbb{N}}$, where F_n/E_n is not generated by a Borel action, such that if $E \subseteq F$ are countable Borel equivalence relations on a Polish space X and F/E is not generated by a Borel action, then there is some n and a continuous embedding of $2^{\mathbb{N}}$ into X that simultaneously embeds E_n to E and F_n to F.

Remark 2.44 In [dRM2], obstructions are also obtained for analogs of the Glimm–Effros dichotomy theorem and the Lusin–Novikov theorem for quotient spaces.

(2) We next consider the ergodicity of actions $G \curvearrowright_B X/E$. Given a pair $E \subseteq F$ as in (1) above, F/E is **ergodic** if there is no Borel partition $X = A \sqcup B$ with A, B E-invariant, complete F-sections. The action $G \curvearrowright_B X/E$ is ergodic if $E^{\vee G}/E$ is ergodic. Ben Miller in [Mi5] showed the following results concerning ergodicity.

For a countable group G, let $F_0(G)$ be the countable Borel equivalence relation on $G^{\mathbb{N}}$ given by

$$(g_0, g_1, g_2, \dots)\ F_0(G)\ (h_0, h_1, h_2, \dots) \iff \exists m\, \forall k > m\, [g_0 \cdots g_k = h_0 \cdots h_k].$$

There is an action $G \curvearrowright_B (G^{\mathbb{N}}, F_0(G))$ defined by

$$g \cdot (g_0, g_1, g_2, \dots) = (g \cdot g_0, g_1, g_2, \dots),$$

inducing an ergodic action $G \curvearrowright_B G^{\mathbb{N}}/F_0(G)$. Let E be a countable Borel equivalence relation on a Polish space X, and let $G \curvearrowright_B X/E$ be a free action. Then $E^{\vee G}/E$ is ergodic if and only if there is a G-equivariant Borel injection of $G^{\mathbb{N}}/F_0(G)$ to X/E induced by a continuous embedding of $G^{\mathbb{N}}$ to X (see [Mi5, Chapter 2, Theorem 7.2]). If $E^{\vee G}$ is hyperfinite, then there is a G-equivariant Borel injection of X/E to $G^{\mathbb{N}}/F_0(G)$ (see [Mi5, Chapter 2, Theorem 8.1]).

(3) In [FKS], the lifting problem for actions on quotients is studied. Let E be a countable Borel equivalence relation on X. Denote by p_E the canonical surjective homomorphism of $\mathrm{Aut}_B(E)$ onto $\mathrm{Out}_B(E)$. There is also a canonical homomorphism of $\mathrm{Aut}_B(E)$ into $\mathrm{Sym}_B(X/E)$ that sends $T \in \mathrm{Aut}_B(E)$ to the permutation $[x]_E \mapsto [T(x)]_E$. This has kernel $\mathrm{Inn}_B(E)$, and therefore it gives an injective homomorphism i_E from $\mathrm{Out}_B(E)$ into $\mathrm{Sym}_B(X/E)$. Its range consists of the so-called **outer permutations**. An action $G \curvearrowright_B (X, E)$, which can be viewed as a homomorphism from G into $\mathrm{Aut}_B(E)$, gives via p_E a homomorphism from G into $\mathrm{Out}_B(E)$ and finally through i_E to a homomorphism from G to $\mathrm{Sym}_B(X/E)$, i.e., an action $G \curvearrowright_B X/E$. One considers here the question of when a homomorphism of G to $\mathrm{Sym}_B(X/E)$ lifts to a homomorphism of G to $\mathrm{Aut}_B(E)$, i.e., when an action $G \curvearrowright_B X/E$ lifts to an action $G \curvearrowright_B (X, E)$. The first result is the following:

Theorem 2.45 ([FKS, Theorem 1.1]) *Let E be a compressible countable Borel equivalence relation. Then every Borel action $G \curvearrowright_B X/E$ has a lift $G \curvearrowright_B (X, E)$.*

This has the following corollary:

Corollary 2.46 ([FKS, Corollary 1.2]) *Let E be an aperiodic countable Borel equivalence relation on a Polish space X. Then for any action $G \curvearrowright_B X/E$, there is a comeager $E^{\vee G}$-invariant Borel subset $Y \subseteq X$ such that $G \curvearrowright_B Y/E$ lifts.*

There are examples of E for which not every element of $\mathrm{Sym}_B(X/E)$ is outer, and therefore there are actions $G \curvearrowright_B X/E$, i.e., homomorphisms of G into $\mathrm{Sym}_B(X/E)$, which do not even lift to homomorphisms of G into $\mathrm{Out}_B(E)$ and thus much less to actions $G \curvearrowright_B (X, E)$.

Call then an action $G \curvearrowright_B X/E$ **outer** if it acts by outer permutations. The next question is then to find out when an *outer* action $G \curvearrowright_B X/E$ lifts to an action $G \curvearrowright_B (X, E)$. Here we have the following result:

Theorem 2.47 ([FKS, Theorem 1.3]) *Outer actions of amenable groups and amalgamated free products of finite groups have a lift.*

However, there is a limit to the groups G that can have the lifting property for outer actions. In the following, a countable group G is called **treeable** if it admits a free Borel action on a standard Borel space X that has an invariant probability Borel measure (see Chapter 4) and for which the associated orbit equivalence relation E_G^X is treeable (see Section 9.1). For example, every amenable group and every free group is treeable, but infinite property (T) groups and products of an infinite group and a nonamenable group are not treeable.

Proposition 2.48 ([FKS, Proposition 1.4]) *If every outer action of a countable group G lifts, then G is treeable.*

Thus if $\mathcal{G}$ is the class of countable groups that have the lifting property for outer actions, then $\mathcal{G}$ contains all the amenable groups and all amalgamated free products of finite groups, and it is closed under subgroups (see [FKS, Proposition 4.7]) and free products. Finally, all groups in $\mathcal{G}$ are treeable.

It is not known whether one can characterize the class of groups that have this lifting property.

2.12 The Borel Classes of Countable Borel Equivalence Relations

Lecomte [Lec] characterizes when a countable Borel equivalence relation belongs to the Borel classes $\mathbf{\Sigma}^0_\xi$ or $\mathbf{\Pi}^0_\xi$ ($\xi \geq 1$) in terms of a dichotomy. (In fact he proves a more general result for Borel equivalence relations with F_σ classes.)

In what follows, we state his result for $\xi \geq 3$, in which case the dichotomy is somewhat easier to state. For each Polish space X and $\mathbf{\Gamma}$ one of the classes $\mathbf{\Sigma}^0_\xi$ or $\mathbf{\Pi}^0_\xi$ ($\xi \geq 1$), let $\mathbf{\Gamma}(X)$ be the class of subsets of X in $\mathbf{\Gamma}$. We also let $\check{\mathbf{\Gamma}}$ be the dual class of $\mathbf{\Gamma}$, i.e., $\check{\mathbf{\Gamma}}$ is $\mathbf{\Pi}^0_\xi$ or $\mathbf{\Sigma}^0_\xi$, resp. Let $H = 2 \times 2^{\mathbb{N}}$, and for $\xi \geq 3$ and $\mathbf{\Gamma}$ as above, let $C \subseteq 2^{\mathbb{N}}$ be such that $C \cap N_s \in \check{\mathbf{\Gamma}}(N_s) \setminus \mathbf{\Gamma}$, for each $s \in \bigcup_n 2^n$. (Such C are shown to exist.) Define the equivalence relation $E_3^{\mathbf{\Gamma}}$ on H by

$$(i,x) E_3^{\mathbf{\Gamma}} (j,y) \iff (i,x) = (j,y) \text{ or } (x = y \in C).$$

Clearly $E_3^{\mathbf{\Gamma}}$ is a countable Borel equivalence relation and $E_3^{\mathbf{\Gamma}} \in \check{\mathbf{\Gamma}}(H^2) \setminus \mathbf{\Gamma}$. Then we have:

Theorem 2.49 ([Lec, Theorem 1.4]) *Let $\mathbf{\Gamma}$ be one of the classes $\mathbf{\Sigma}^0_\xi$ or $\mathbf{\Pi}^0_\xi$, for $\xi \geq 3$; let X be a Polish space; and let E be a countable Borel equivalence relation on X. Then exactly one of the following holds:*

(i) $E \in \mathbf{\Gamma}(X^2)$.
(ii) $E_3^{\mathbf{\Gamma}} \sqsubseteq_c E$.

3

Essentially Countable Relations

3.1 Essentially Countable and Reducible-to-Countable Equivalence Relations

It turns out that the scope of the theory of countable Borel equivalence relations is much wider as it encompasses many other classes of equivalence relations up to Borel bireducibility.

Definition 3.1 A Borel equivalence relation E is **essentially countable** if it is Borel bireducible to a countable Borel equivalence relation. It is called **reducible-to-countable** if it is Borel reducible to a countable Borel equivalence relation.

Remark 3.2 In the literature, the term "essentially countable" is often used for what we here call "reducible-to-countable."

These two notions are distinct.

Theorem 3.3 ([H4]) *There is a Borel equivalence relation that is reducible-to-countable but not essentially countable.*

For many naturally occurring Borel equivalence relations, the notions coincide. To explain this we need a definition first.

Definition 3.4 Let E be a Borel equivalence relation on a standard Borel space X. Then E is **idealistic** if there is a map $C \in X/E \mapsto \mathcal{I}_C$, assigning to each E-class C a σ-ideal $\mathcal{I}_C$ of subsets of C, with $C \notin \mathcal{I}_C$, such that $C \mapsto \mathcal{I}_C$ is Borel in the following (weak) sense: For each Borel set $A \subseteq Y \times X$, Y a standard Borel space, the set $A_{\mathcal{I}} \subseteq Y \times X$ defined by $(y,x) \in A_{\mathcal{I}} \iff \{x' \in [x]_E : (y,x') \in A\} \in \mathcal{I}_{[x]_E}$ is Borel.

A typical example of an idealistic E is a Borel equivalence relation induced by a Borel action of a Polish group (see [Ke2, page 285]). In the next result we will also need the following definition.

Definition 3.5 Let E be an equivalence relation on a space X. A **complete countable section** for E is a complete section S such that for each $x \in X$, $S \cap [x]_E$ is countable.

If a Borel equivalence relation E admits a complete countable Borel section, it is clearly essentially countable. Now we have the following theorem.

Theorem 3.6 *Let E be an idealistic Borel equivalence relation. Then the following are equivalent:*

(i) *E is reducible-to-countable.*
(ii) *E is essentially countable.*
(iii) *E admits a complete countable Borel section.*

In fact if $E \leq_B F$, where F is a countable Borel equivalence relation on a standard Borel space Y, then there is an F-invariant Borel set $B \subseteq Y$ such that $E \sim_B F \upharpoonright B$.

For a proof, see the more general result in [KMa, 3.7, 3.8] (and the corrections posted in: `pma.divisions.caltech.edu/people/alexander-kechris`).

The following characterization of reducibility to countable is often useful in establishing this property.

Proposition 3.7 (Kechris) *Let E be a Borel equivalence relation on a standard Borel space X. Then the following are equivalent:*

(i) *E is reducible-to-countable.*
(ii) *There is a standard Borel space Y and a Borel function $f\colon X \to Y$ such that (a) $f([x]_E)$ is countable, $\forall x \in X$; and (b) $\neg(xEy) \implies f([x]_E) \cap f([y]_E) = \emptyset$.*

A proof can be found in [Ka, 7.6.1].

Hjorth has proved a dichotomy for the property of essential countability of Borel equivalence relations induced by Borel actions of Polish groups. Let G be a Polish group and $a\colon G \times X \to X$ a continuous action of G on a Polish space X. This action is **stormy** if for every nonempty open set $U \subseteq G$ and any $x \in X$, the map $g \mapsto g \cdot x$ from U to $U \cdot x$ in not an open map.

We now have the following theorem.

Theorem 3.8 ([H5, Theorem 1.3]) *Let G be a Polish group and $a\colon G \times X \to X$ a Borel action of G on a standard Borel space X. Then if the associated equivalence relation E_a is Borel, exactly one of the following holds:*

(i) *E_a is reducible-to-countable.*
(ii) *There is a stormy action b of G on a Polish space Y and a Borel embedding of the action b to the action a, i.e., a Borel injection $F: Y \to X$ such that $F(g \cdot y) = g \cdot F(y)$.*

Moreover, if the action a on a Polish space X is continuous, F can be taken to be continuous too.

We next proceed to discuss various classes of essentially countable Borel equivalence relations.

3.2 Actions of Locally Compact Groups and Lacunary Sections

A rich source of essentially countable Borel equivalence relations comes from actions of locally compact groups. Before we state the main result here, we introduce some additional concepts.

Definition 3.9 Let G be a topological group, and let $a: G \times X \to X$ be an action of G on a space X. A complete section S of E_a is **lacunary** if there is a neighborhood U of the identity of G such that for all $s \in S$, $U \cdot s \cap S = \{s\}$. In this case, we also say that S is ***U*-lacunary**. A complete section S is called **cocompact** if there is a compact neighborhood U of the identity of G such that $U \cdot S = X$. Again in this case we say that S is ***U*-cocompact**.

Note that if G is second countable, then any lacunary complete section is countable.

If a Polish locally compact group G acts in a Borel way on a standard Borel space X, the induced equivalence relation is Borel (use, for example, [Ke6, 9.17 and 35.46]). We now have the following theorem.

Theorem 3.10 ([Ke2]) *Let G be a Polish locally compact group, and let $a: G \times X \to X$ be a Borel action of G on a standard Borel space X. For any compact neighborhood U of the identity of G, E_a has a complete U-lacunary Borel section S. In particular E_a is essentially countable.*

A measure-theoretic version of this result (where null sets are neglected) was proved in [FHM] (and for the free action case in [Fo]), and the case $G = \mathbb{R}$ of Theorem 3.10 was proved in [Wa] (while the measure-theoretic version in the case of $\mathbb{R}$ was proved in [Am]).

Other proofs of Theorem 3.10 can be found in [Ke8, pages 244–245], [H5, 2.2] and [HMT, 7.8].

One can also make the lacunary sections to be cocompact. This was shown independently by Conley and Dufloux (unpublished).

Theorem 3.11

(i) (Conley, Dufloux) *Let G be a Polish locally compact group, and let $a\colon G \times X \to X$ be a Borel action of G on a standard Borel space X. Then E_a has a complete lacunary cocompact Borel section.*
(ii) [Sl1, 2.4] *In fact, let U be a symmetric compact neighborhood of the identity of G, and put $V = U^2$. Then for any complete V-lacunary Borel section S of E_a, there is a maximal (under inclusion) complete V-lacunary Borel section $T \supseteq S$ of E_a, and thus T is V-cocompact.*

A measure-theoretic version of such a result for free actions can also be found in [KPV, 4.2].

One can formulate these results in the language of descriptive combinatorics. Let G be a Polish locally compact group, and let $a\colon G \times X \to X$ be a Borel action of G on a standard Borel space X. Let U be a symmetric compact neighborhood of the identity of G, and put $V = U^2$. Define the Borel graph whose set of vertices is X and distinct $x, y \in X$ are connected by an edge if and only if $y \in V \cdot x$. Then Theorems 3.10 and 3.11 imply that in this graph there exists a maximal independent set that is Borel.

The following gives a more detailed analysis of the equivalence relation induced by a locally compact group action.

Theorem 3.12 ([Ke4, Theorem 1]) *Let G be a Polish locally compact group, and let $a\colon G \times X \to X$ be a Borel action of G on a standard Borel space X. Then there is a (unique) decomposition $X = A \sqcup B$ of X into invariant Borel sets such that $E_a \restriction A$ is countable and $E_a \restriction B \cong_B F \times I_{\mathbb{R}}$, where $F = E_a \restriction S$, with S a countable complete Borel section of $E_a \restriction B$.*

Theorem 3.13 ([Ke4, Theorem 2]) *The map $E \mapsto E \times I_{\mathbb{R}}$ induces a bijection between countable Borel equivalence relations, up to Borel bireducibility, and equivalence relations induced by Borel actions of Polish locally compact groups with uncountable orbits, up to Borel isomorphism.*

One interesting application of Theorem 3.10 is in the proof of the result in [HK2] that isomorphism (conformal equivalence) of Riemann surfaces, and in particular complex domains (open connected subsets of $\mathbb{C}$), is an essentially countable Borel equivalence relation. We will discuss this in more detail in Section 11.2.5).

Concerning lacunary sections as in Theorem 3.10, it is of interest in certain situations to obtain additional information about their structure. For the case of free Borel actions of $\mathbb{R}$ on standard Borel spaces, each orbit is an (affine) copy of $\mathbb{R}$, so if S is a complete lacunary section, it makes sense to talk about the distance between consecutive members of S in the same orbit. We now have the following result that provides a purely Borel strengthening of a classical result of Rudolph [Ru] in the measure-theoretic context and again neglecting null sets.

Theorem 3.14 ([Sl2]) *Let α, β be two rationally independent positive reals. Then any free Borel action of $\mathbb{R}$ admits a complete lacunary Borel section such that the distance between any two consecutive points in the same orbit belongs to $\{\alpha, \beta\}$.*

The paper [Sl3] further provides, for a set of positive reals A and a free Borel action of $\mathbb{R}$, criteria for the existence of a complete lacunary Borel section such that the distance between any two consecutive points in the same orbit belongs to A. The papers [Sl1] and [Sl4] use complete lacunary and cocompact sections to prove classification results for $\mathbb{R}^n$-actions.

In [Mi13] the following generalization of lacunarity was introduced. A Borel action a of a Polish group on a standard Borel space X is called **σ-lacunary** if X can be decomposed into countably many invariant Borel sets on each of which the induced action admits a complete lacunary Borel section. Clearly for any such action, in which E_a is Borel, E_a is essentially countable. In fact the converse holds as well.

Theorem 3.15 ([Gr1]) *Let G be a Polish group, and let a be a Borel action of G on a standard Borel space with E_a Borel. Then the following are equivalent:*

(i) *The action is σ-lacunary.*
(ii) *E_a is essentially countable.*

Finally we consider the question of whether Theorem 3.10 actually characterizes Polish locally compact groups.

Problem 3.16 Let G be a Polish group with the property that all the equivalence relations induced by Borel actions of G on standard Borel spaces are Borel and essentially countable. Is the group locally compact?

In [Tho] it is shown that every such group must be CLI, i.e., admit a complete left-invariant metric. However there are many CLI groups that are not locally compact.

An affirmative answer to Problem 3.16 has been obtained for certain classes of Polish groups:

(i) [So] all separable Banach spaces, viewed as groups under addition;

(ii) [Ma] all *abelian* isometry groups of separable locally compact metric spaces;

(iii) [KMPZ] all isometry groups of separable locally compact metric spaces. This class of groups includes those in (ii) and all non-Archimedean Polish groups.

A related result characterizing Polish compact groups was also proved in [So]: A Polish group is compact if and only if all the equivalence relations induced by Borel actions of G on standard Borel spaces are Borel and smooth.

3.3 On the Existence of Complete Countable Borel Sections

We note here that the existence of a complete countable Borel section for a Borel equivalence relation is in general stronger than being essentially countable. The standard example is as follows: Let $X \subseteq \mathbb{R} \times \mathbb{R}$ be a Borel set that projects onto $\mathbb{R}$ but admits no Borel uniformization (see, e.g., [Ke6, 18.17]). Let E on X be defined by $(x, y)E(x', y') \iff x = x'$. Then $E \sim_B \Delta_{\mathbb{R}}$, but E admits no complete countable Borel section. The following result gives a characterization of the existence of complete countable Borel sections.

We call a Borel equivalence relation E **ccc idealistic** if it satisfies Definition 3.4 with the σ-ideals $\mathcal{I}_C$ satisfying in addition the countable chain condition (ccc; i.e., any pairwise disjoint collection of subsets of C not in $\mathcal{I}_C$ is countable). For example, a Borel equivalence relation induced by a Borel action of a Polish group is ccc idealistic.

In the following, for σ-finite Borel measures μ, ν, on the same standard Borel space, $\mu \sim \nu$ denotes **measure equivalence**, i.e., having the same null sets. Finally we call a Borel equivalence relation E **σ-smooth** if it can be written as $E = \bigcup_n E_n$, where each E_n is a smooth Borel equivalence relation.

Theorem 3.17 ([Ke2, 1.5]) *Let E be a Borel equivalence relation on a standard Borel space X. Then the following are equivalent:*

(i) *E admits a complete countable Borel section.*

(ii) (a) *E is σ-smooth; and* (b) *E is ccc idealistic.*

(iii) *As in* (ii) *but with* (b) *replaced by* (b)*: *There is a Borel assignment $x \mapsto \mu_x$ of probability Borel measures to points $x \in X$ such that $\mu_x([x]_E) = 1$ and $xEy \implies \mu_x \sim \mu_y$.*

A generalization of the measure-theoretic result in [FHM] is proved in [R]. It states that if E is a Borel equivalence relation on a standard Borel space X, μ is a probability Borel measure on X, and there is a Borel assignment $x \mapsto \mu_x$ of probability Borel measures to points $x \in X$ such that $\mu_x([x]_E) = 1$ and

$xEy \implies \mu_x \sim \mu_y$, then E admits a complete countable Borel section, μ-a.e.. It is unknown if there is a purely Borel version of this result, i.e., whether condition (a) is necessary in Theorem 3.17(iii).

Problem 3.18 Is it true that the following are equivalent?

(i) E admits a complete countable Borel section.
(ii) There is a Borel assignment $x \mapsto \mu_x$ of probability Borel measures to points $x \in X$ such that $\mu_x([x]_E) = 1$ and $xEy \implies \mu_x \sim \mu_y$.

It is known that condition (a) is necessary in Theorem 3.17(ii); see the discussion in [Ke2, Section 1, (III), (IV)].

Note that if E is a Borel equivalence relation that is reducible-to-countable, then, in Theorem 3.17(ii), (a) is automatically satisfied by Corollary 2.7, so for such equivalence relations the existence of a countable complete Borel section is equivalent to condition (b) and also to condition (b)* and, by Theorem 3.6, also to the condition that E is idealistic.

Another characterization has been found in [H10]. In the following, a Borel equivalence relation E on a standard Borel space X is called **treeable** if there is a Borel acyclic graph on X such that the E-classes are exactly its connected components. It is called **σ-treeable** if it can be written as $E = \bigcup_n E_n$, where each E_n is Borel and treeable. We now have the following theorem.

Theorem 3.19 ([H10]) *Let E be a Borel equivalence relation that is reducible-to-countable. Then the following are equivalent:*

(i) *E admits a complete countable Borel section.*
(ii) *E is σ-treeable.*

Moreover, the following holds, which shows that, in Theorem 3.17(ii), condition (a) can be replaced by σ-treeability.

Theorem 3.20 ([H10]) *Let E be a Borel equivalence relation on a standard Borel space X. Then the following are equivalent:*

(i) *E admits a complete countable Borel section.*
(ii) (a) *E is σ-treeable; and* (b) *E is ccc idealistic.*

It is also shown in [H10] that every σ-treeable smooth Borel equivalence relation admits a Borel transversal.

Finally, in [CLM], dichotomy theorems are proved characterizing when a treeable Borel equivalence relation admits a complete countable Borel section and also when the equivalence relation E_T, for a Borel function T, admits a complete countable Borel section.

3.4 Actions of Non-Archimedean Polish Groups

Recall that a Polish group is **non-Archimedean** if it has a neighborhood basis at the identity consisting of open subgroups. Equivalently these are (up to topological group isomorphism) the closed subgroups of the infinite symmetric group S_∞ of all permutations of $\mathbb{N}$, with the pointwise convergence topology, and also the automorphism groups of countable structures (in the sense of model theory); see [BK]. It turns out that one can characterize exactly which Borel actions of such groups induce Borel equivalence relations that are essentially countable. To formulate this we need the following definition.

Definition 3.21 Let E be a Borel equivalence relation on a standard Borel space and Λ a class of sets in Polish spaces, closed under continuous preimages. Then E is **potentially Λ** if there is an equivalence relation F, in some Polish space, which is in the class Λ, such that $E \leq_B F$. This is equivalent to saying that there is a Polish topology τ on X inducing its Borel structure such that E is in the class Λ (in the product space $(X^2, \tau \times \tau)$).

For example, it turns out that a Borel equivalence relation is smooth if and only if it is potentially $\mathbf{\Pi}^0_1$ if and only if it is potentially $\mathbf{\Pi}^0_2$ (see [HKL]). We now have the following theorem.

Theorem 3.22 ([HK1, 3.8], [HKLo, 4.1]) *Let G be a Polish non-Archimedean group, and let $a\colon G \times X \to X$ be a Borel action of G on a standard Borel space X. Then the following are equivalent:*

(i) *E_a is essentially countable.*
(ii) *E_a is potentially $\mathbf{\Sigma}^0_2$.*
(iii) *E_a is potentially $\mathbf{\Sigma}^0_3$.*

This fails for arbitrary Polish groups (see, e.g., [HK1, first remark in page 236]) and also for Polish non-Archimedean groups if $\mathbf{\Sigma}^0_3$ is replaced by $\mathbf{\Pi}^0_3$ (see [HKLo]; a specific example is the equivalence relation E_{ctble}, on the G_δ subspace of injective sequences in $\mathbb{R}^{\mathbb{N}}$, given by $(x_n) E_{\text{ctble}} (y_n) \iff \{x_n : n \in \mathbb{N}\} = \{y_n : n \in \mathbb{N}\}$, which is $\mathbf{\Pi}^0_3$ but not essentially countable).

3.5 Logic Actions and the Isomorphism Relation on the Countable Models of a Theory

Fix a countable relational language $L = \{R_i\}_{i \in I}$, where R_i has arity n_i. We denote by $X_L = \text{Mod}_{\mathbb{N}}(L)$ the space of L-structures with universe $\mathbb{N}$. Thus X_L can be identified with the compact metrizable space $\prod_{i \in I} 2^{\mathbb{N}^{n_i}}$. We let $\mathbb{A} \cong \mathbb{B}$

be the isomorphism relation between structures in X_L. This is the equivalence relation induced by the so-called **logic action** of the infinite symmetric group S_∞ of all permutations of $\mathbb{N}$ on X_L, given by $g \cdot \mathbb{A} = \mathbb{B}$ if and only if g is an isomorphism of $\mathbb{A}$ with $\mathbb{B}$. For each sentence $\sigma \in L_{\omega_1\omega}$, let

$$\mathrm{Mod}(\sigma) = \{\mathbb{A} \in X_L : \mathbb{A} \models \sigma\}$$

be the set of models of σ. This is a Borel invariant under isomorphism subset of X_L, and, by the classical theorem of Lopez-Escobar, every such Borel subset of X_L is of the form $\mathrm{Mod}(\sigma)$, see [Ke6, 16.8]. Denote by $\cong_\sigma$ the isomorphism relation restricted to $\mathrm{Mod}(\sigma)$. It is shown in [BK, 2.7.3] that if $a\colon S_\infty \times X \to X$ is a Borel action of S_∞ on a standard Borel space X, then a is Borel isomorphic to the logic action on some $\mathrm{Mod}(\sigma)$.

Remark 3.23 In cases where we want to consider languages with function symbols, we will replace them by their graphs.

We next state model-theoretic criteria for essential countability (and smoothness) of $\cong_\sigma$. Let F be a countable fragment of $L_{\omega_1\omega}$, see [B]. For any L-structure $\mathbb{A}$, we denote by $\mathrm{Th}_F(\mathbb{A})$ the set of sentences in F that hold in $\mathbb{A}$. We say that a countable L-structure $\mathbb{A}$ is **$\aleph_0$-categorical for** F if, for every countable L-structure $\mathbb{B}$ for which $\mathrm{Th}_F(\mathbb{A}) = \mathrm{Th}_F(\mathbb{B})$, $\mathbb{B}$ is isomorphic to $\mathbb{A}$. We now have the following theorem.

Theorem 3.24 ([HK1, 4.2]) *Let $\sigma \in L_{\omega_1\omega}$. Then the following are equivalent:*

(i) *$\cong_\sigma$ is Borel and smooth.*
(ii) *There is a countable fragment F of $L_{\omega_1\omega}$ containing σ, such that every countable model $\mathbb{A}$ of σ is $\aleph_0$-categorical for F.*

Theorem 3.25 ([HK1, 4.3]) *Let $\sigma \in L_{\omega_1\omega}$. Then the following are equivalent:*

(i) *$\cong_\sigma$ is Borel and essentially countable.*
(ii) *There is a countable fragment F of $L_{\omega_1\omega}$ containing σ, such that for every countable model $\mathbb{A} = \langle A, \dots\rangle$ of σ, there is an $n \geq 1$ and a finite sequence $\bar{a} \in A^n$ such that $\langle \mathbb{A}, \bar{a}\rangle$ is $\aleph_0$-categorical for F.*

Using this last result, one can easily prove the essential countability of the isomorphism relation on the following structures: finitely generated groups (or, more generally, finitely generated structures in some countable language), connected locally finite graphs, locally finite trees, finite transcendence degree over $\mathbb{Q}$ fields, torsion-free abelian groups of finite rank, etc.

Remark 3.26 (i) For the case of torsion-free abelian groups of finite rank $\leq n$, one can also directly see that the isomorphism relation is essentially countable, since it is Borel bireducible to the equivalence relation on the space of subgroups of $\langle \mathbb{Q}^n, + \rangle$ induced by the action of $\mathrm{GL}_n(\mathbb{Q})$ on this space.

(ii) Also in the case of finitely generated groups, one can also directly see that the isomorphism relation is essentially countable by using the space of finitely generated groups; see, e.g., [T8, Section 1].

We conclude with the following open problem of Hjorth and Kechris.

Problem 3.27 Let σ be a *first-order* theory, i.e., the conjunction of countably many first-order sentences. Is it possible for $\cong_\sigma$ to be Borel, nonsmooth and essentially countable?

A negative answer has been obtained in [Mar] for first-order theories with uncountably many types.

3.6 Another Example

Let $\mathbb{U}$ be the **Urysohn metric space** and $F(\mathbb{U})$ the standard Borel space of closed subsets of $\mathbb{U}$ with the Effros Borel structure. We view $F(\mathbb{U})$ as the standard Borel space of Polish metric spaces. Let $\mathcal{M}$ be a class of Polish metric spaces closed under isometry. We call $\mathcal{M}$ a **Borel** class if $\mathcal{M} \cap F(\mathbb{U})$ is Borel in $F(\mathbb{U})$. Denote by $\cong^{iso}$ the equivalence relation of isometry on $F(\mathbb{U})$, and let $\cong^{iso}_{\mathcal{M}}$ be its restriction to $\mathcal{M} \cap F(\mathbb{U})$, i.e, the equivalence relation of isometry for spaces in $\mathcal{M}$. See [GK] for the study of this equivalence relation on various classes of Polish metric spaces. Recall that a metric space is **proper** (or **Heine–Borel**) if every closed bounded set is compact.

Theorem 3.28 (Hjorth; see [GK, 7.1]) *There is a Borel class $\mathcal{M}$ of Polish metric spaces, closed under isometry, such that $\mathcal{M}$ contains all connected locally compact Polish metric spaces and all proper locally compact Polish metric spaces and such that $\cong^{iso}_{\mathcal{M}}$ is an essentially countable Borel equivalence relation.*

3.7 Dichotomies Involving Reducibility to Countable

For the subsequent theorems we use the following terminology and notation. The equivalence relation E_2 on $2^{\mathbb{N}}$ is defined by

$$xE_2y \iff \sum_{\{n\in\mathbb{N}:\, x_n\neq y_n\}} \frac{1}{n+1} < \infty.$$

It can be shown that E_2 is Borel bireducible to the equivalence relation induced by the translation action of $\ell^1 = \{(x_n) \in \mathbb{R}^{\mathbb{N}} \colon \sum_n |x_n| < \infty\}$ on $\mathbb{R}^{\mathbb{N}}$, see [Ka, 6.2.4]. We also let

$$E_3 = E_0^{\mathbb{N}}.$$

A Polish group is **tsi** if it admits a compatible two-sided invariant metric. For non-Archimedean groups, this is equivalent to admitting a neighborhood basis at the identity consisting of normal open subgroups.

We now have the following dichotomy theorems.

Theorem 3.29 ([H2, 0.4]) *Let E be a Borel equivalence relation on a standard Borel space. If $E \le_B E_2$, then exactly one of the following holds:*

(i) *E is reducible-to-countable.*

(ii) *$E_2 \le_B E$.*

Theorem 3.30 ([HK3, 8.1]) *Let E be a Borel equivalence relation on a standard Borel space. If $E \le_B E_a$, where a is a Borel action of a non-Archimedean tsi Polish group G with E_a Borel, then exactly one of the following holds:*

(i) *E is reducible-to-countable.*

(ii) *$E_3 \le_B E$.*

In [Gr1, 1.2] it is shown that for $E = E_a$, alternative (i) in Theorem 3.30 is equivalent to the following statement:

(i′) There is a sequence (N_n) of open normal subgroups of G and for each n a continuous action a_n of G/N_n on a Polish space such that $E_a \le_B \bigoplus_n E_{a_n}$.

Also, for generalizations of Theorem 3.30, again in the case $E = E_a$, for *arbitrary* tsi Polish groups, see [Mi13, 4.1] and [Gr2, 1.4].

Finally, the following dichotomy is established in [dRM2], where E_1 is the following equivalence relation on $\mathbb{R}^{\mathbb{N}}$

$$(x_n) E_1 (y_n) \iff \exists m \forall n \ge m (x_n = y_n).$$

Theorem 3.31 ([dRM2]) *Let E be a σ-smooth Borel equivalence relation on a Polish space. Then exactly one of the following holds:*

(i) *E is reducible-to-countable.*

(ii) *$E_1 \sqsubseteq_c E$.*

3.8 Canonization

In the book [KSZ], the authors study canonization theorems, which analyze the behavior of equivalence relations on “large sets.” For example, Silver’s Theorem 5.1 in Chapter 5 implies that if E is a Borel equivalence relation on an uncountable Polish space, E will be trivial on a nonempty perfect set P, i.e., $E \restriction P = \Delta_P$ or $E \restriction P = P^2$. Chapters 4, 7 and 8 in [KSZ] deal with problems of canonization related to countable and reducible-to-countable Borel equivalence relations. For example, a result of Mathias, see [KSZ, 8.17], asserts that for every countable Borel equivalence relation E on the space of all infinite subsets of $\mathbb{N}$, there is an infinite subset $A \subseteq \mathbb{N}$ such that E restricted to the set of all infinite subsets of A is hyperfinite; in fact it is contained in E_0 (where we view E_0 here as the equivalence relation of finite symmetric difference on the space of subsets of $\mathbb{N}$). Another canonization theorem can be found in [PW].

4

Invariant and Quasi-invariant Measures

4.1 Terminology and Notation

For the rest of this survey we adopt the following terminology and notation: A **measure** on a standard Borel space X is a σ-finite Borel measure μ. If $\mu(X) < \infty$, μ is called **finite**, and if $\mu(X) = 1$, it is called a **probability** measure. If μ is a measure on X and $Y \subseteq X$ is a Borel set, then $\mu \upharpoonright Y$ is the measure on Y that is the restriction of μ to the Borel subsets of Y. A measure μ is **absolutely continuous** to a measure ν, in symbols $\mu \ll \nu$, if for every Borel set $A \subseteq X$, $\nu(A) = 0 \implies \mu(A) = 0$, and μ, ν are **equivalent**, in symbols $\mu \sim \nu$, if $\mu \ll \nu$ & $\nu \ll \mu$. The equivalence class of a measure under this equivalence relation is called its **measure class**. Note that every measure is equivalent to a probability measure.

4.2 Invariant Measures

If G is a countable group that acts in a Borel way on a standard Borel space X, then G acts on the set of measures on X by $g \cdot \mu(A) = \mu(g^{-1} \cdot A)$. We say that μ is **invariant** under this action if for all $g \in G$, $g \cdot \mu = \mu$.

Assume now that E is a countable Borel equivalence relation on a standard Borel space X. In the following, denote by $[[E]]_B$ the **full pseudogroup** of E, which is the set of all Borel bijections $f\colon A \to B$ between Borel subsets A, B of X such that $\forall x \in A(f(x)Ex)$. Thus $[E]_B \subseteq [[E]]_B$. Note that if E is induced by a Borel action of a countable group G on X, then $f\colon A \to B$ as above is in $[[E]]_B$ if and only if there are a countable decomposition $A = \bigsqcup_n A_n$ and group elements (g_n) such that for each n and $x \in A_n$, $f(x) = g_n \cdot x$.

Note now the following simple fact, where for a Borel function $T\colon X \to Y$ on standard Borel spaces X, Y and measure μ on X, $T_*\mu$ is the **push-forward measure** on Y, defined by $T_*\mu(B) = \mu(T^{-1}(B))$, for every Borel set $B \subseteq Y$.

Proposition 4.1 *Let E be a countable Borel equivalence relation on a standard Borel space X. Then the following are equivalent for each measure μ on X:*

(i) *For some countable group G and Borel action a of G on X such that $E_a = E$, μ is invariant under this action.*
(ii) *For every countable group G and every Borel action a of G on X such that $E_a = E$, μ is invariant under this action.*
(iii) *For every $f\colon A \to B$ in $[[E]]_B$, $\mu(A) = \mu(B)$.*
(iv) *For every $T \in [E]_B$, $T_*\mu = \mu$.*

Definition 4.2 Let E be a countable Borel equivalence relation on a standard Borel space X and μ be a measure on X. Then μ is called ***E*-invariant** if it satisfies the equivalent conditions of Proposition 4.1. Also μ is ***E*-ergodic** if for every E-invariant Borel set A, we have $\mu(A) = 0$ or $\mu(X \setminus A) = 0$.

In the following, a measure μ is called **nonatomic** if every singleton has measure 0.

Proposition 4.3 (see [DJK, 3.2]) *Let E be a countable Borel equivalence relation on a standard Borel space X and A be a complete Borel section for E. Let ν be a measure on A such that ν is $E \restriction A$-invariant. Then there is a unique E-invariant measure μ on X such that for all Borel sets $B \subseteq A$, $\mu(B) = \nu(B)$. If ν is nonatomic or ergodic, so is μ.*

In particular, if E has an invariant (nonatomic, ergodic) measure and $E \sqsubseteq_B F$, then the same holds for F.

Note that E_0 admits an invariant, nonatomic, ergodic probability measure, namely the usual product measure on $2^{\mathbb{N}}$. Now the general Glimm–Effros dichotomy, proved in [HKL], asserts that a Borel equivalence relation E is not smooth if and only if $E_0 \sqsubseteq_B E$. The special case of this for countable Borel equivalence relations was already proved in [E1], [E2], and [We1]. So we have the following characterization:

Theorem 4.4 ([E1], [E2], [We1]) *Let E be a countable Borel equivalence relation on a standard Borel space X. Then the following are equivalent:*

(i) *E is not smooth.*
(ii) *There is a nonatomic, E-ergodic, E-invariant measure.*

Corollary 4.5 *Every countable Borel equivalence relation on an uncountable standard Borel space admits a nonatomic invariant measure.*

4.3 Invariant Probability Measures and the Theorem of Nadkarni

The following result of Nadkarni characterizes the existence of E-invariant probability measures.

Theorem 4.6 ([N2]; see also [BK, 4.5]) *Let E be a countable Borel equivalence relation on a standard Borel space X. Then the following are equivalent:*

(i) *E is not compressible.*
(ii) *There is an E-invariant probability measure.*
(iii) *There is an E-ergodic, E-invariant probability measure.*

A proof of this result can be also found in [Ke5, 4.G] and [Sl5, 2.3–2.8].

For example, E_t, and eventual equality $E_0(\mathbb{N})$ on $\mathbb{N}^\mathbb{N}$, being compressible, do not admit an invariant probability measure, and E_0, having an invariant probability measure, is not compressible.

Corollary 4.7 *Let E be a countable Borel equivalence relation on a standard Borel space X. Then for any Borel set $A \subseteq X$, the following are equivalent:*

(i) *$[A]_E$ is compressible.*
(ii) *For any E-invariant probability measure μ, $\mu(A) = 0$.*

There is also an effective version of Theorem 4.6 due to Ditzen:

Theorem 4.8 ([Di], Section 2.2, Theorem 1; see also [KWo]) *Let E be a countable Borel equivalence relation on $\mathbb{N}^\mathbb{N}$ that is Δ^1_1 (effectively Borel). Then the following are equivalent:*

(i) *There is no Δ^1_1 compression of E.*
(ii) *There is an E-invariant probability measure.*

It follows from Theorem 2.29 that for every aperiodic countable Borel equivalence relation E on a Polish space X, there is a comeager invariant Borel set C such that $E \upharpoonright C$ admits no invariant probability measure. For a related result involving stationary measures, see [CKM, Corollary 18].

Remark 4.9 When G is a unimodular Polish locally compact group, a is a free Borel action of G on a standard Borel space X, and $Y \subseteq X$ is a complete lacunary cocompact Borel section, then there are a canonical correspondence between finite invariant measures for $E_a \upharpoonright Y$ and finite invariant measures for the action a; see [Sl1, Section 4] and [KPV, Section 4].

4.4 Ergodic Invariant Measures and the Ergodic Decomposition Theorem

For each standard Borel space X, denote by $P(X)$ the standard Borel space of probability measures on X, which is generated by the maps $\mu \mapsto \mu(A)$, for Borel sets A; see [Ke6, 17.23, 17.24]. For a countable Borel equivalence relation E, denote by INV_E the set of E-invariant probability measures. Also let EINV_E be the set of E-ergodic, E-invariant probability measures. The following is an important property of such measures:

Proposition 4.10 *Let E be a countable Borel equivalence relation on a standard Borel space X, and let $A, B \subseteq X$ be Borel sets. Then there is $f \in [[E]]_B, f : C \to D$, such that $C \subseteq A, D \subseteq B$, and $\mu(A \setminus C) = 0$, $\forall \mu \in \mathrm{EINV}_E$ with $\mu(A) \leq \mu(B)$.*

A proof can be found, for example, in [KM1, proof of 7.10].

The next results apply as well to Borel actions of Polish locally compact groups, but we will restrict attention here to actions of countable groups or equivalently to countable Borel equivalence relations.

Theorem 4.11 ([Fa], [Va]) *Let E be a countable Borel equivalence relation on a standard Borel space X. Then*

(i) INV_E, EINV_E *are Borel sets in $P(X)$, and* EINV_E *is the set of the extreme points of the convex set* INV_E *(under the usual operation of convex combination of probability measures).*
(ii) $\mathrm{INV}_E \neq \emptyset \iff \mathrm{EINV}_E \neq \emptyset$.

The following result is known as the **ergodic decomposition theorem** for invariant measures.

Theorem 4.12 ([Fa], [Va]) *Let E be a countable Borel equivalence relation on a standard Borel space X, and assume that* $\mathrm{INV}_E \neq \emptyset$. *Then there is a Borel surjection $\pi : X \to \mathrm{EINV}_E$ such that:*

(i) *π is E-invariant.*
(ii) *If $X_e = \pi^{-1}(\{e\})$, for $e \in \mathrm{EINV}_E$, then $e(X_e) = 1$ and e is the unique E-invariant probability measure concentrating on X_e.*
(iii) *If $\mu \in \mathrm{INV}_E$, then $\mu = \int \pi(x)\, \mathrm{d}\mu(x) = \int e\, \mathrm{d}\pi_*\mu(e)$.*

Moreover, this map is unique in the following sense: If π, π' satisfy (i)–(iii), then for any $\mu \in \mathrm{INV}_E$, $\pi(x) = \pi'(x)$, μ-a.e. (x) (equivalently by Corollary 4.7, the set $\{x : \pi(x) \neq \pi'(x)\}$ is compressible).

Corollary 4.13 *Let E be a countable Borel equivalence relation on a standard Borel space X, and let $\mu, \nu \in \mathrm{INV}_E$. Then $\mu = \nu$ if and only if for every E-invariant Borel set $A \subseteq X$, $\mu(A) = \nu(A)$.*

Proofs of Theorems 4.11 and 4.12 can be also found in [Ke5, 3.K], [S15, 2.9], and [Ch3]. Moreover, [Di, Section 2.1.2, Theorem 10] (see also [KWo]) gives an effective version of Theorem 4.12.

In [Ch2] a connection is found between the ergodic decomposition of a countable Borel equivalence relation E on a standard Borel space X and the topological ergodic decomposition of continuous (in Polish topologies on X that generate its Borel structure) actions of countable groups G that generate E. Here the **topological ergodic decomposition** of an action of a group G on a topological space X is the equivalence relation on X, where two points of X are equivalent if the closures of their orbits coincide.

4.5 Quasi-invariant Measures

If G is a countable group that acts in a Borel way on a standard Borel space X and μ is a measure on X, then μ is **quasi-invariant** under this action if for all $g \in G$, $g \cdot \mu \sim \mu$. Note that if μ is quasi-invariant under the action and $\nu \sim \mu$, then ν is also quasi-invariant under this action. Thus for every quasi-invariant measure there is an equivalent quasi-invariant probability measure.

We now have the following analog of Proposition 4.1.

Proposition 4.14 *Let E be a countable Borel equivalence relation on a standard Borel space X. Then the following are equivalent for each measure μ on X:*

(i) *For some countable group G and Borel action a of G on X such that $E_a = E$, μ is quasi-invariant under this action.*
(ii) *For every countable group G and every Borel action a of G on X such that $E_a = E$, μ is quasi-invariant under this action.*
(iii) *For every $f : A \to B$ in $[[E]]_B$, $\mu(A) = 0 \iff \mu(B) = 0$.*
(iv) *For every $T \in [E]_B$, $T_*\mu \sim \mu$.*
(v) *For every Borel $A \subseteq X$, $\mu(A) = 0 \iff \mu([A]_E) = 0$.*

Definition 4.15 Let E be a countable Borel equivalence relation on a standard Borel space X and μ be a measure on X. Then μ is called ***E*-quasi-invariant** if it satisfies the equivalent conditions of Proposition 4.14.

The following two results show that studying the structure of countable Borel equivalence relations with respect to arbitrary measures can often be reduced to that of quasi-invariant measures.

Proposition 4.16 *Let E be a countable Borel equivalence relation on a standard Borel space X and μ be a measure on X. Then there is an E-quasi-invariant measure $\bar{\mu}$ such that:*

(i) *$\mu \ll \bar{\mu}$, and if ν is E-quasi-invariant with $\mu \ll \nu$, then $\bar{\mu} \ll \nu$.*
(ii) *For any Borel E-invariant set $A \subseteq X$, $\mu(A) = \bar{\mu}(A)$.*
(iii) *If μ is nonatomic or ergodic, so is $\bar{\mu}$.*

Proof Let E be generated by a Borel action of a countable group $G = \{g_n\}$ and put $\bar{\mu} = \sum_n 2^{-n-1}(g_n \cdot \mu)$. □

Proposition 4.17 (Woodin, see [HK1, 6.5] and [Mi14, 1.3]) *Let E be a countable Borel equivalence relation on a standard Borel space X, and let μ be a probability measure on X. Then there is complete Borel section $Y \subseteq X$ with $\mu(Y) = 1$ such that if $A \subseteq Y$ is a Borel set with $\mu(A) = 0$, then $\mu([A]_E) = 0$. Therefore $\mu \upharpoonright Y$ is $E \upharpoonright Y$-quasi-invariant.*

The following is an analog of Proposition 4.3:

Proposition 4.18 (see [DJK, 3.3]) *Let E be a countable Borel equivalence relation on a standard Borel space X and A be a complete Borel section for E. Let ν be a probability measure on A such that ν is $E \upharpoonright A$-quasi-invariant. Then there is an E-quasi-invariant probability measure μ on X such that for all Borel sets $B \subseteq A$, $\mu(B) = \mu(A)\nu(B)$. If ν is nonatomic or ergodic, so is μ.*

And the following is an analog of Theorem 4.4:

Theorem 4.19 ([E1], [E2], [We1]) *Let E be a countable Borel equivalence relation on a standard Borel space X. Then the following are equivalent:*

(i) *E is not smooth.*
(ii) *There is a nonatomic, E-ergodic, E-quasi-invariant measure.*

The following classical results of Hopf and Hajian–Kakutani characterize the existence of an invariant probability measure equivalent to a given quasi-invariant measure (compare with Theorem 4.6). In the following, if a countable group G acts in a Borel way on a standard Borel space X, a **weakly wandering** Borel set for this action is a Borel set $A \subseteq X$ such that for some sequence (g_n) of elements of G, we have $g_n \cdot A \cap g_m \cdot A = \emptyset$, for all $m \neq n$.

Theorem 4.20 (Hopf, see [N3, Section 10]; Hajian–Kakutani, see [HaK]) *Let E be a countable Borel equivalence relation on a standard Borel space X and μ be an E-quasi-invariant measure. Then the following are equivalent:*

(i) *There is an E-invariant probability measure ν such that $\mu \sim \nu$.*
(ii) (Hopf) *There is no Borel set A with $\mu(A) > 0$ such that $E \upharpoonright A$ is compressible.*
(iii) (Hajian–Kakutani) *Let G be a countable group and let a be a Borel action of G on X such that $E_a = E$. Then there is no weakly wandering set of μ-positive measure for this action.*

Other related characterizations can be found in [Ke5, 2.83–2.85].

4.6 The Space of Quasi-invariant Measures

Let E be a countable Borel equivalence relation on a standard Borel space X. We denote by QINV_E, EQINV_E, and ERG_E, the spaces of E-quasi-invariant, E-ergodic and E-quasi-invariant E-ergodic probability measures on X, resp. The following is a special case of a more general result concerning Borel actions of Polish locally compact groups proved in [Di].

Theorem 4.21 ([Di]) *Let E be a countable Borel equivalence relation on a standard Borel space X. Then* QINV_E, EQINV_E, and ERG_E *are Borel sets in* $P(X)$.

The set EQINV_E is invariant under measure equivalence $\sim$, which is a Borel equivalence relation, and [DJK, 4.1] shows that if E is not smooth, measure equivalence on EQINV_E (even restricted to nonatomic measures) is not smooth.

The structure of QINV_E and EQINV_E under absolute continuity depends only on the bireducibility type of E.

Proposition 4.22 ([DJK, 4.2]) *Let E, F be countable Borel equivalence relations on standard Borel spaces. If $E \sim_B F$, then there are Borel maps* $\varphi\colon \mathrm{QINV}_E \to \mathrm{QINV}_F$ *and* $\psi\colon \mathrm{QINV}_F \to \mathrm{QINV}_E$ *such that:*

(i) $\psi(\varphi(\mu)) \sim \mu$ *and* $\varphi(\psi(\nu)) \sim \nu$.
(ii) $\mu \ll \nu \iff \varphi(\mu) \ll \varphi(\nu)$.
(iii) φ, ψ *map ergodic measures to ergodic measures.*

We also have the following related fact:

Proposition 4.23 ([Ke5, 4.34]) *Let E, F be countable Borel equivalence relations on standard Borel spaces such that $E \sqsubseteq_B F$. Then there is a Borel injection φ: $\mathrm{QINV}_E \to \mathrm{QINV}_F$ such that $\mu \ll \nu \iff \varphi(\mu) \ll \varphi(\nu)$ and φ maps ergodic measures to ergodic measures.*

The paper [Ke12] studies the descriptive complexity of the Borel sets $\mathrm{QINV}_\Gamma = \mathrm{QINV}_E$ and $\mathrm{EQINV}_\Gamma = \mathrm{EQINV}_E$, where E is the equivalence relation induced by the shift action of a countable group Γ on $X = (2^{\mathbb{N}})^\Gamma$. The following is shown, where the sets therein belong to the space $P(X)$, which is given the usual compact metrizable topology (see, e.g., [Ke6, Section 17.E]) that generates its standard Borel structure. Also $\mathbb{F}_\infty$ is the free group with a countably infinite set of generators.

Theorem 4.24 ([Ke12])

(i) *For each infinite countable group Γ, the set QINV_Γ is $\mathbf{\Pi}^0_3$-complete and EQINV_Γ is $\mathbf{\Pi}^0_3$-hard.*
(ii) *The set $\mathrm{EQINV}_\mathbb{Z}$ is $\mathbf{\Pi}^0_3$-complete.*
(iii) *There is a countable ordinal $3 \leq \alpha_\infty \leq \omega + 2$ such that $\mathrm{EQINV}_{\mathbb{F}_\infty}$ is $\mathbf{\Pi}^0_{\alpha_\infty}$-complete.*

The exact value of α_∞ is unknown.

4.7 The Cocycle of a Quasi-invariant Measure

Let E be a countable Borel equivalence relation on a standard Borel space X, and let μ be a probability measure. We define two Borel measures M_l, M_r on the space E (viewed as a Borel subset of X^2) as follows:

$$M_l(A) = \int |A_x| \, \mathrm{d}\mu(x),$$

where $A \subseteq E$ is Borel, $A_x = \{y \in X : (x, y) \in A\}$, and $|B|$ is the cardinality of B, which is equal to ∞ if B is infinite. For any nonnegative real-valued Borel φ,

$$\int \varphi(x, y) \, \mathrm{d}M_l(x, y) = \int \sum_{y \in [x]_E} \varphi(x, y) \mathrm{d}\mu(x).$$

Let also

$$M_r(A) = \int |A^y| \, \mathrm{d}\mu(y),$$

where $A^y = \{x \in X : (x, y) \in A\}$. Note that for $f\colon A \to B$ in $[[E]]_B$,

$$M_l(\mathrm{graph}(f)) = \mu(A), M_r(\mathrm{graph}(f)) = \mu(B);$$

thus clearly M_l, M_r are σ-finite. Moreover, μ is E-quasi-invariant if and only if $M_l \sim M_r$, and μ is E-invariant if and only if $M_r = M_l$.

Assume now that μ is E-quasi-invariant. Consider then the Radon–Nikodym derivative,

$$\rho_\mu(x, y) = (dM_l/dM_r)(x, y),$$

for $(x, y) \in E$. Therefore ρ_μ is a Borel map from E to $\mathbb{R}^+$ such that for any nonnegative real-valued $\varphi\colon E \to \mathbb{R}$, we have

$$\int \varphi(x, y)\, \mathrm{d}M_l(x, y) = \int \varphi(x, y)\rho_\mu(x, y)\, \mathrm{d}M_r(x, y);$$

thus for every Borel set $A \subseteq E$,

$$M_l(A) = \int_A \rho_\mu(x, y)\, \mathrm{d}M_r(x, y),$$

and ρ_μ is uniquely determined M_r-a.e. by this property. Moreover $\rho_\mu^{-1} = \mathrm{d}M_r/\mathrm{d}M_l$, M_l-a.e.

For any $f\colon A \to B$ in $[[E]]_B$ and Borel $C \subseteq B$, we have

$$\mu(f^{-1}(C)) = \int_C \rho_\mu(f^{-1}(y), y)\, \mathrm{d}\mu(y);$$

thus if $T \in [E]_B$, then $(\mathrm{d}T_*\mu/\mathrm{d}\mu)(x) = \rho_\mu(T^{-1}(x), x)$, μ-a.e. (x).

A map $\rho : E \to G$, where G is a group, is called a **cocycle** if it satisfies the **cocycle identity**

$$\rho(x, z) = \rho(y, z)\rho(x, y)$$

for $xEyEz$. If this cocycle identity holds only on an E-invariant Borel set A with $\mu(A) = 1$, then ρ is a **cocycle a.e.** We now have that $\rho_\mu : E \to \mathbb{R}^+$ (the multiplicative group of positive reals) is a cocycle a.e. We thus call ρ_μ the **cocycle associated** with the E-quasi-invariant measure μ.

Proofs of the facts mentioned here can be found in [KM1, Section 8].

4.8 Existence of Quasi-invariant Probability Measures with a Given Cocycle

Again we will see that generically there are no quasi-invariant probability measures with a given cocycle (compare with the paragraph following Theorem 4.8). For a countable Borel equivalence relation E on a standard Borel space

X and $\rho\colon E \to \mathbb{R}^+$ a Borel cocycle, we say that E is **ρ-aperiodic** if for every $x \in X$, $\sum_{y\in[x]_E} \rho(y,x) = \infty$. Also we say that a probability measure μ on X is **ρ-invariant** if it is E-quasi-invariant and $\rho_\mu(x,y) = \rho(x,y)$, for all xEy in an E-invariant Borel set A with $\mu(A) = 1$. For more information about such measures, see [Mi2, Section 18] and [Mi14].

Theorem 4.25 ([KM1, 13.1]) *Let E be a countable Borel equivalence relation on a Polish space X, and let $\rho\colon E \to \mathbb{R}^+$ be a Borel cocycle such that E is ρ-aperiodic. Then there is an E-invariant comeager Borel set $C \subseteq X$ such that $\mu(C) = 0$, for any ρ-invariant probability measure μ.*

In the papers [Mi2, Section 20], [Mi9], [Mi12], and [Mi14], Miller obtains analogs of Nadkarni's Theorem 4.6 by characterizing when, for a given countable Borel equivalence relation E on a standard Borel space X and Borel cocycle $\rho\colon E \to \mathbb{R}^+$, there exists a ρ-invariant probability measure. Analogous results for measures (as opposed to probability measures) were obtained in [Mi2, Section 18], [Mi8], and [Mi14].

4.9 An Ergodic Decomposition Theorem for Quasi-invariant Measures with a Given Cocycle

The following is a generalization of the ergodic decomposition theorem (Theorem 4.12) to measures having a given cocycle (Theorem 4.12 is the special case of the constant value-1 cocycle).

Let E be a countable Borel equivalence relation on a standard Borel space X. For each Borel cocycle $\rho\colon E \to \mathbb{R}^+$, let INV_ρ, resp., EINV_ρ, be the spaces of ρ-invariant, resp., E-ergodic, ρ-invariant probability measures on X.

Theorem 4.26 ([Di, Section 2.1, Theorem 6]) *Let E be a countable Borel equivalence relation on a standard Borel space X and $\rho\colon E \to \mathbb{R}^+$ be a Borel cocycle. Then:*

(a) INV_ρ, EINV_ρ *are Borel sets in $P(X)$, and* EINV_ρ *is the set of the extreme points of the convex set* INV_ρ *(under the usual operation of convex combination of probability measures).*
(b) $\mathrm{INV}_\rho \neq \emptyset \iff \mathrm{EINV}_\rho \neq \emptyset$.

Moreover, there is a Borel surjection $\pi\colon X \to \mathrm{EINV}_\rho$ such that:

(i) *π is E-invariant.*
(ii) *If $X_e = \pi^{-1}(\{e\})$, for $e \in \mathrm{EINV}_\rho$, then $e(X_e) = 1$ and e is the unique ρ-invariant measure concentrating on X_e.*

(iii) *If* $\mu \in \mathrm{INV}_\rho$, *then* $\mu = \int \pi(x)\, \mathrm{d}\mu(x) = \int e\, \mathrm{d}\pi_*\mu(e)$.

Moreover, this map is unique in the following sense: If π, π' *satisfy (i)–(iii), then for any* $\mu \in \mathrm{INV}_\rho$, $\pi(x) = \pi'(x)$, μ*-a.e.* (x).

Another proof of this result is given in [Mi9, 5.2]. Also [Di, Section 2.1, Theorem 10] gives an effective version of Theorem 4.26.

4.10 An Ergodic Decomposition Theorem with respect to an Arbitrary Probability Measure

The following result is a special case of a more general theorem concerning analytic equivalence relations, see [Ke5, 3.J] and [LM, 3.2].

Proposition 4.27 (Kechris) *Let* E *be a countable Borel equivalence relation on a standard Borel space* X, *and let* $\mu \in P(X)$. *Then there is a Borel* E*-invariant map* $\pi \colon X \to P(X)$ *such that letting* $\pi(x) = \mu_x$, *we have:*

(i) μ_x *is* E*-ergodic,* μ*-a.e.* (x).
(ii) $\{y \in X \colon \mu_y = \mu_x\}$ *has* μ_x*-measure 1.*
(iii) $\mu = \int \mu_x\, \mathrm{d}\mu(x)$.

4.11 Measures Agreeing on Invariant Sets

Recall from Corollary 4.13 that if E is a countable Borel equivalence relation and $\mu, \nu \in \mathrm{INV}_E$, then μ, ν agree on the E-invariant Borel sets if and only if $\mu = \nu$. In fact it turns out that one can characterize exactly when two *arbitrary* probability measures agree on the E-invariant Borel sets. The following result is due to [Th, Theorem 1], where it is proved more generally for equivalence relations induced by Borel actions of Polish locally compact groups. See also [Kh, Section 3.4].

Theorem 4.28 ([Th, Theorem 1]) *Let* a *be a Borel action of a countable group* G *on a standard Borel space* X, *and let* $E = E_a$. *Let* μ, ν *be two probability measures on* X. *Then the following are equivalent:*

(i) *For each* E*-invariant Borel set* $A \subseteq X$, $\mu(A) = \nu(A)$.
(ii) *There is a probability measure* ρ *on* E *such that* $s_*\rho = \mu, t_*\rho = \nu$, *where* $s(x, y) = x, t(x, y) = y$.
(iii) *There is a probability measure* σ *on* $G \times X$ *such that* $u_*\sigma = \mu, v_*\sigma = \nu$, *where* $u(g, x) = x, v(g, x) = g \cdot x$.

(iv) *There is a Borel map* $x \mapsto \mu_x$ *from* X *to* $P(G)$ *such that for every Borel set* $A \subseteq X$, $\nu(A) = \int \mu_x(\{g \in G : g \cdot x \in A\})\, \mathrm{d}\mu(x)$.

(v) *There is a map* $g \mapsto \mu_g$ *from* G *to the set of measures on* X *such that* $\mu = \sum_g \mu_g$ *and* $\nu = \sum_g g \cdot \mu_g$.

In [Sh], this is generalized to the context of cardinal algebras, in particular also leading to an algebraic proof of this result.

5

Smoothness, E_0 and E_∞

We will now start studying the hierarchical order $\leq_B$ of Borel reducibility on countable Borel equivalence relations.

5.1 Smoothness

The simplest Borel equivalence relations are the smooth ones, and they are easy to classify up to Borel bireducibility. In the following, for $n \geq 1$, Δ_n is the equality relation on a set of cardinality n.

First we recall the following dichotomy result of Silver:

Theorem 5.1 ([Si]) *Let E be a Borel (or even $\mathbf{\Pi}^1_1$) equivalence relation on a standard Borel space X. Then exactly one of the following holds:*

(i) *There are only countably many E-classes.*
(ii) $\Delta_\mathbb{R} \sqsubseteq_B E$.

We now have as an immediate consequence:

Corollary 5.2 *If E is a smooth Borel equivalence relation, then $E \sim_B \Delta_n$, for some $n \geq 1$, $E \sim_B \Delta_\mathbb{N}$ or $E \sim_B \Delta_\mathbb{R}$.*

Moreover, the smooth countable Borel equivalence relations form an initial segment in $\leq_B$.

Corollary 5.3 *We have that*

$$\Delta_1 <_B \Delta_2 <_B \cdots <_B \Delta_n <_B \cdots <_B \Delta_\mathbb{N} <_B \Delta_\mathbb{R},$$

and every Borel equivalence E is either Borel bireducible to an equivalence relation in this list or else $\Delta_\mathbb{R} <_B E$.

Remark 5.4 Given a pair $E_1 \subseteq E_2$ of countable Borel equivalence relations on a standard Borel space X and a pair $F_1 \subseteq F_2$ of countable Borel equivalence relations on a standard Borel space Y, we say that (E_1, E_2) is **simultaneously Borel reducible** to (F_1, F_2) if there is a Borel function $f\colon X \to Y$ such that $f\colon E_1 \leq_B F_1, f\colon E_2 \leq_B F_2$. The paper [Sc1] contains a classification of pairs $E_1 \subseteq E_2$ of smooth countable Borel equivalence relations up to simultaneous Borel bireducibility.

5.2 The Simplest Nonsmooth Relation

There is a least complicated, in the sense of Borel reducibility, Borel equivalence relation $>_B \Delta_{\mathbb{R}}$.

Theorem 5.5 (The general Glimm–Effros dichotomy, [HKL]) *Let E be a Borel equivalence relation on a standard Borel space. Then exactly one of the following holds:*

(i) *E is smooth.*
(ii) *$E_0 \sqsubseteq_B E$.*

We note that for the case of a countable Borel equivalence relation E, this result is already included in [E1], [E2], and [We1].

We can thus extend the initial segment given in Corollary 5.3.

Theorem 5.6 *We have that*

$$\Delta_1 <_B \Delta_2 <_B \cdots <_B \Delta_n <_B \cdots <_B \Delta_{\mathbb{N}} <_B \Delta_{\mathbb{R}} <_B E_0,$$

and every Borel equivalence E is either Borel bireducible to an equivalence relation in this list or else $E_0 <_B E$.

The countable Borel equivalence relations that are $\leq_B E_0$ are exactly the hyperfinite ones and will be studied in detail in Chapter 7.

5.3 The Most Complicated Relation

At the other end of the spectrum there is a most complicated, in terms of Borel reducibility, countable Borel equivalence relation.

For each countable group G and standard Borel space X, denote by $s_{G,X}$ the **shift action** of G on the space X^G:

$$(g \cdot p)_h = p_{g^{-1}h}$$

for $p \in X^G$, $g, h \in G$. Let $E(G, X) = E_{s_{G,X}}$ be the associated equivalence relation. In the following, let $\mathbb{F}_n$, $n = 1, 2, \ldots$, be the **free group** with n generators. We also let $\mathbb{F}_\infty$ be the free group with a countably infinite set of generators.

If a is a Borel action of a countable group G on a standard Borel space X, which we can assume is a Borel subset of $\mathbb{R}$, then the map

$$x \in X \mapsto (g^{-1} \cdot x)_{g \in G} \in \mathbb{R}^G$$

is a Borel embedding of the action a to the shift action $s_{G,\mathbb{R}}$, so in particular $E_a \sqsubseteq_B^i E(G, \mathbb{R})$.

Definition 5.7 A countable Borel equivalence relation E is **invariantly universal** if for every countable Borel equivalence relation F, $F \sqsubseteq_B^i E$.

Since by Theorem 2.3 every countable Borel equivalence relation is induced by a Borel action of $\mathbb{F}_\infty$, it follows that the equivalence relation $E(\mathbb{F}_\infty, \mathbb{R})$ is invariantly universal. Clearly there is a unique, up to Borel isomorphism, invariantly universal countable Borel equivalence relation, and it will be denoted by E_∞.

Definition 5.8 We say that a countable Borel equivalence relation E is **universal** if for any countable Borel equivalence relation F, we have $F \leq_B E$, i.e., $E \sim_B E_\infty$.

Another example of a universal countable Borel equivalence relation is the following:

Proposition 5.9 ([DJK, 1.8]) $E(\mathbb{F}_2, 2) \sim_B E_\infty$.

In [T15], S. Thomas studies what he calls E_0-extensions, i.e., countable Borel equivalence relations of the form $E(a)$, where E, on some space X, is Borel isomorphic to E_0 and $a\colon G \curvearrowright_B (X, E)$. Up to Borel isomorphism these are exactly the countable Borel equivalence relations that are normal over E_0. For consistency with our terminology, we will call these $\boldsymbol{E_0}$**-expansions**. He shows in [T15, Theorem 1.2] that for $\mathbb{F}_2$, and in fact any countable group G containing $\mathbb{F}_2$, the following E_0-expansion is universal: Let E be the analog of E_0 on the space 2^G (i.e., $pEq \iff \{g\colon p(g) \neq q(g)\}$ is finite), so that $E \cong_B E_0$. Let $s = s_{G,2}$ be the shift action of G. Clearly this action is by automorphisms of E. Then the E_0-expansion $E(s)$ is universal.

We will study universal countable Borel equivalence relations in Chapter 11.

5.4 Intermediate Relations

The interval $[E_0, E_\infty]$ in the Borel reducibility order $\leq_B$ is not trivial, as one can prove, for example, using the results in [A1] and [SlSt]. In fact we have the following, where by $F(G, X)$ we denote the restriction of $E(G, X)$ to the (invariant Borel) **free part** $F(X^G) = \{p \in X^G : \forall g \neq 1_G(g \cdot p \neq p)\}$ of the shift action $s_{G,X}$.

Theorem 5.10 ([JKL, Section 3]) $E_0 <_B F(\mathbb{F}_2, 2) <_B E_\infty$.

The relation $F(\mathbb{F}_2, 2)$ is an example of a treeable countable Borel equivalence relation, a concept that we will study in detail in Chapter 9.

We call countable Borel equivalence relations E such that $E_0 <_B E <_B E_\infty$ **intermediate**. In the next chapter, Chapter 6, we will see that they contain many interesting examples and have a very rich structure.

It should be pointed out that all known proofs of existence of intermediate countable Borel equivalence relations use measure-theoretic methods of ergodic theory. We will see in Chapter 7 that generically, in the sense of Baire category, all countable Borel equivalence relations are $\leq_B E_0$.

6

Rigidity and Incomparability

6.1 The Complex Structure of Borel Reducibility

By the early 1990s, a small finite number of intermediate countable Borel equivalence relations were known, and they were linearly ordered under $\leq_B$. This led to the following basic problems: Are there infinitely many, up to Borel bireducibility? Does nonlinearity occur here?

These problems were resolved in [AK], where it was shown that the structure of countable Borel equivalence relations under Borel reducibility is quite rich.

Theorem 6.1 ([AK, Theorem 1]) *The partial order of Borel sets under inclusion can be embedded into the quasi-order of Borel reducibility of countable Borel equivalence relations, i.e., there is a map $A \mapsto E_A$ from the Borel subsets of $\mathbb{R}$ to countable Borel equivalence relations such that $A \subseteq B \iff E_A \leq_B E_B$.*

In particular it follows that any Borel partial order can be embedded into the quasi-order of Borel reducibility of countable Borel equivalence relations. Under the **Continuum Hypothesis (CH)**, any partial order of the size of the continuum can be embedded into the partial order of inclusion of subsets of $\mathbb{N}$ modulo finite sets. It follows that for every quasi-order $\leq$ on a set X of the size of the continuum, there is a map $x \mapsto E_x$ from X to countable Borel equivalence relations such that $x \leq y \iff E_x \leq_B E_y$, i.e., $\leq_B$ on countable Borel equivalence relations is a universal quasi-order of the size of the continuum.

Other proofs of Theorem 6.1 can be found in [HK4], [CM1], and [T12].

Another indication of the complexity of $\leq_B$ on countable Borel equivalence relations is contained in the next result. To formulate it, fix a coding (parametrization) of countable Borel equivalence relation by reals. This consists of a $\mathbf{\Pi}^1_1$ subset C of $\mathbb{R}$ and a surjective map $c \mapsto E_c$ from C to the set of countable Borel equivalence relations on $\mathbb{R}$, satisfying some natural definability conditions; see [AK, Section 5]. Let

$$C_{\leq} = \{(c,d) \in C^2 \colon E_c \leq_B E_d\}, C_{\sim} = \{(c,d) \in C^2 \colon E_c \sim_B E_d\},$$

$$C_{\cong} = \{(c,d) \in C^2 \colon E_c \cong_B E_d\}.$$

Theorem 6.2 ([AK, Theorem 2]) *The sets $C_{\leq}, C_{\sim}, C_{\cong}$ are $\mathbf{\Sigma}^1_2$-complete.*

Concerning the equivalence relation $C_{\sim}$, it follows from Theorem 6.1 that every Borel equivalence relation can be Borel reduced to $C_{\sim}$, and in [G3] this was extended to $\mathbf{\Sigma}^1_1$ equivalence relations. It appears to be unknown if it also holds for $\mathbf{\Pi}^1_1$ equivalence relations.

The proof of Theorem 6.1 used methods of ergodic theory, more precisely Zimmer's cocycle superrigidity theory for ergodic actions of linear algebraic groups and their lattices.

The key point is that there is a phenomenon of set-theoretic rigidity analogous to the measure-theoretic rigidity phenomena discovered by Zimmer; see [Z2]. Informally this can be described as follows:

- **(Measure-theoretic rigidity)** Under certain circumstances, when a countable group acts preserving a probability measure, the equivalence relation associated with the action together with the measure "encode" or "remember" significant information about the group (and the action).
- **(Borel-theoretic rigidity)** Such information is simply encoded in the Borel cardinality of the (quotient) orbit space.

6.2 Cocycle Reduction

A basic idea in establishing rigidity results in the measure-theoretic, as well as in the descriptive, context is cocycle reduction, which we will discuss next in the framework of Borel reducibility.

Let E be a countable Borel equivalence relation on a standard Borel space X and Γ be a countable group. Two Borel cocycles $\alpha \colon E \to \Gamma, \beta \colon E \to \Gamma$ are **cohomologous** or **equivalent**, in symbols $\alpha \sim \beta$, if there is a Borel map $f \colon X \to \Gamma$ such that $xEy \implies \beta(x,y) = f(y)\alpha(x,y)f(x)^{-1}$. If G is a countable group acting in a Borel way on a standard Borel space X, a Borel cocycle of this action into Γ is a Borel map $\alpha \colon G \times X \to \Gamma$ such that $\alpha(g_1 g_2, x) = \alpha(g_1, g_2 \cdot x)\alpha(g_2, x)$, for $g_1, g_2 \in G$. Two such cocycles α, β are equivalent, in symbols $\alpha \sim \beta$, if there is a Borel map $f \colon X \to \Gamma$ such that $\beta(g,x) = f(g \cdot x)\alpha(g,x)f(x)^{-1}$. Note that if α is a cocycle of the equivalence relation E associated to this action, then α gives a cocycle β of the action, namely $\beta(g,x) = \alpha(x, g \cdot x)$. Conversely if the action is free, any cocycle of the action gives rise as above to a cocycle of the equivalence relation.

A simple example of a cocycle of an action of G on X is a homomorphism $h\colon G \to \Gamma$, which can be identified with the cocycle $\alpha(g,x) = h(g)$.

Now let F be a countable Borel equivalence relation on a standard Borel space Y, which is induced by a *free* Borel action b of a countable group Γ on Y, i.e., $F = E_b$. Let $\varphi\colon E \to_B F$ be a Borel homomorphism. Then there is a canonical Borel cocycle $\alpha\colon E \to \Gamma$ associated to φ, namely $\alpha(x,y) = \gamma$, where γ is the unique element of Γ such that $\gamma \cdot \varphi(x) = \varphi(y)$, i.e., $xEy \implies \varphi(y) = \alpha(x,y) \cdot \varphi(x)$. Note now that if $a \sim \beta$, via the Borel function $f\colon X \to \Gamma$, then β is also associated to a Borel homomorphism $\psi\colon E \to_B F$, such that $\varphi(x) F \psi(x)$ for every $x \in X$, namely $\psi(x) = f(x) \cdot \varphi(x)$. Similarly, if $E = E_a$ for a Borel action a of a countable group G on X, we have an associated to φ cocycle of the action, given by $\varphi(g \cdot x) = \alpha(g,x) \cdot \varphi(x)$.

A cocycle reduction result, for some given action or equivalence relation, shows that certain cocycles of the action or equivalence relation are equivalent to ones that are much simpler in some sense, e.g., are group homomorphisms or have a "small range." When such a cocycle reduction result is applied to the cocycle coming from a homomorphism of equivalence relations as above, it can be used to replace the given homomorphism with another one that has additional structure.

For example, let a be a Borel action of a countable group G on a standard Borel space X and b be a free Borel action of a countable group Γ on a standard Borel space Y. Put $E = E_a, F = E_b$. Let $\varphi\colon E \to_B F$ be a Borel homomorphism, and let $\alpha\colon G \times X \to \Gamma$ be the cocycle associated to φ of the action a as above. If α is equivalent to a homomorphism $h\colon G \to \Gamma$, let ψ be the homomorphism of E to F associated to the cocycle h as above. Then we have that $\psi(g \cdot x) = h(g) \cdot \psi(x)$, which is a very strong property that can be ruled out in a given situation, thereby ruling out the existence of the original homomorphism φ. This therefore gives a basic technique for showing that one equivalence relation cannot be reduced to another.

In practice such cocycle reduction results are actually established in a measure-theoretic context, i.e., in ergodic theory. Suppose we have a countable Borel equivalence relation E on a standard Borel space X with an invariant probability measure μ. Then we can define the above notion of cocycle for E, μ by requiring that the cocycle identity holds only on an E-invariant Borel set of μ-measure 1, i.e., we consider cocycles a.e. Moreover, we identify two such cocycles if they agree μ-a.e. Analogously we define equivalence of cocycles, again neglecting μ-null sets. We can also similarly define cocycles of group actions with an invariant measure.

To illustrate these ideas, let us mention a cocycle reduction result due to Popa [Po], which is usually referred to as **cocycle superrigidity**.

Theorem 6.3 ([Po]) *Let G be an infinite countable group with property* (T), *and consider the shift action $s = s_{G,[0,1]}$ of G on $[0,1]^G$ with the invariant product measure λ^G, where λ is Lebesgue measure on $[0,1]$. Then for any countable group Γ, any Borel cocycle of this action into Γ is equivalent, a.e., to a homomorphism of G into Γ.*

Such cocycle reduction results are used to prove measure-theoretic rigidity results in the sense of Section 6.1. For example, here is an application of Theorem 6.3. In what follows, if a is a Borel action of a countable group G on a standard Borel space X with invariant probability measure μ, and if b is a Borel action of a countable group Γ on a standard Borel space Y with invariant probability measure ν, we say that a, b are **orbit equivalent** if there are invariant Borel sets $X_0 \subseteq X, Y_0 \subseteq Y$ with $\mu(X_0) = 1, \nu(Y_0) = 1$ and a measure-preserving Borel bijection $T\colon X_0 \to Y_0$, such that $T\colon E_a|X_0 \cong_B E_b|Y_0$. We now have:

Theorem 6.4 ([Po]) *Let G be a simple infinite countable group with property* (T), *and let $s = s_{G,[0,1]}$ be its shift action on $X = [0,1]^G$ with the product measure $\mu = \lambda^G$. Let Γ be a countable group, and let a be a free Borel action of Γ on a standard Borel space Y with invariant probability measure ν. If s, a are orbit equivalent, then there are an isomorphism $h\colon G \to \Gamma$ and a Borel isomorphism $T\colon (X,\mu) \to (Y,\nu)$ such that $T(g \cdot x) = h(g) \cdot T(x)$, μ-a.e. (x).*

Thus in Theorem 6.4, the equivalence relation E_a and the measure μ determine completely the group and the action.

See also [Ke10, Section 30, (B)] for an exposition of the proofs of Theorems 6.3 and 6.4.

In the Borel context, a method that has been frequently employed to solve the problem of showing that one countable Borel equivalence relation cannot be Borel reducible to another ultimately comes down to an application of a cocycle reduction theorem in a measure-theoretic context, usually after considerable technical work that often employs sophisticated methods of ergodic theory or other subjects depending on the context. Such a technique has been used in the proof of Theorem 6.1 and also in all the results that will be mentioned later in this section.

Other methods, concerning the problem of showing failure of Borel reducibility, have been used, e.g., in [HK4, Chapters 6,7 and Appendices B3, B4] (based on earlier work of Furstenberg, Zimmer, and Adams), Hjorth [H3], [H12], [Ke9], [ET], [Mi11], [CM1], and [CM2].

In the rest of this section, we will discuss concrete natural instances of Borel-theoretic rigidity phenomena that allow us to distinguish up to Borel bireducibility countable Borel equivalence relations or to show that one cannot be Borel reducible to another.

6.3 Actions of Linear Groups

The proof of Theorem 6.1 was based on the following result, which uses cocycle reduction results of Zimmer; see [Z2] and references contained therein.

Definition 6.5 If E is a Borel equivalence relation on a standard Borel space X, μ is a probability measure on X, and F is a Borel equivalence relation on a standard Borel space Y, we say that E is $\boldsymbol{\mu, F}$**-ergodic** if for any $f\colon E \to_B F$, there is a Borel E-invariant set $A \subseteq X$, with $\mu(A) = 1$, such that f maps A into a single F-class.

Theorem 6.6 ([AK, 4.5]) *For each nonempty set of primes S, consider the group $G_S = \mathrm{SO}_7(\mathbb{Z}[S^{-1}])$ of 7×7 orthogonal matrices with determinant 1 with coefficients in the ring $\mathbb{Z}[S^{-1}]$ of rationals whose denominators, in reduced form, have prime factors in S. Then*

$$S \not\subseteq T \implies F(G_S, 2) \text{ is } \mu_S, F(G_T, 2)\text{–ergodic},$$

where μ_S is the usual product measure on 2^{G_S}.

In particular,

$$S \subseteq T \iff F(G_S, 2) \leq_B F(G_T, 2).$$

Another set-theoretic rigidity result proved in [AK] is the following:

Theorem 6.7 ([AK, Section 7, (i)]) *Consider the canonical action of* $\mathrm{GL}_n(\mathbb{Z})$ *on $\mathbb{R}^n/\mathbb{Z}^n$, and let G_n be the associated countable Borel equivalence relation. Then*

$$m < n \iff G_m <_B G_n.$$

In particular, the Borel cardinality of the orbit space of this action "encodes" the dimension n.

Fix an integer $n > 1$. For each nonempty set of primes S, consider the compact group $H^n_S = \prod_{p \in S} \mathrm{SL}_n(\mathbb{Z}_p)$. The group $\mathrm{SL}_n(\mathbb{Z})$ can be viewed as a (dense) subgroup of H^n_S via the diagonal embedding. Denote by E^n_S the equivalence relation on H^n_S induced by the translation action of $\mathrm{SL}_n(\mathbb{Z})$ on H^n_S. Then we have:

Theorem 6.8 ([T4, Theorem 5.1], for $n \geq 3$; [I2, Corollary C] for $n = 2$)

$$S = T \iff E^n_S \leq_B E^n_T.$$

Next, for each prime p and $n \geq 2$, consider the projective space $\mathrm{PG}(n-1, \mathbb{Q}_p)$ over the field $\mathbb{Q}_p$ of p-adic numbers, i.e., the space of 1-dimensional vector subspaces of the n-dimensional vector space $\mathbb{Q}^n_p$. Then the group $\mathrm{GL}_n(\mathbb{Z})$ acts in the usual way on $\mathrm{PG}(n-1, \mathbb{Q}_p)$. Denote by F^n_p the associated countable Borel equivalence relation. Then we have:

Theorem 6.9 ([T4, Theorem 6.7], for $n \geq 3$; [I2, Corollary D] for $n = 2$)

$$p = q \iff F^n_p \leq_B F^n_q.$$

In [I2] further such set-theoretic rigidity results are proved for similar actions of nonamenable subgroups of $\mathrm{SL}_2(\mathbb{Z})$.

For a finite set S of primes, a prime number p and $n \geq 2$, denote by $F^n_{p,S}$ the countable Borel equivalence relation induced by the action of $\mathrm{SL}_n(\mathbb{Z}[S^{-1}])$ on $\mathrm{PG}(n-1, \mathbb{Q}_p)$. Also, for a set of primes J such that $S \cap J = \emptyset$, view $\mathrm{SL}_n(\mathbb{Z}[S^{-1}])$ as a subgroup of $\prod_{p \in J} \mathrm{SL}_n(\mathbb{Z}_p)$, and let $F^n_{S,J}$ be the associated equivalence relation induced by the translation action. Then we have:

Theorem 6.10 ([T6, Theorem 1.1, Theorem 1.2],[T4]) *Let $n \geq 2$.*

(i) *If p, q are primes and S, T are nonempty finite sets of primes with $p \notin S, q \notin T$, then*

$$(p, S) = (q, T) \iff F^n_{p,S} \leq_B F^n_{q,T}.$$

(ii) *If S, T are finite nonempty sets of primes and J, K are nonempty sets of primes with $S \cap J = \emptyset, T \cap K = \emptyset$, then*

$$(S, J) = (T, K) \iff F^n_{S,J} \leq_B F^n_{T,K}.$$

For $n = 2$ this result is contained in [T6], while for $n \geq 3$ it is implicit in [T4].

For other set-theoretic rigidity results for actions of linear groups as above, see also [T1], [Co2], [Co4], [I2], and [C].

6.4 Actions of Product Groups

There are also set-theoretic rigidity results concerning actions of product groups.

Theorem 6.11 ([HK4, Theorem 1]) *For each nonempty set S of odd primes, let $H_S = (\star_{p\in S}(\mathbb{Z}/p\mathbb{Z} \star \mathbb{Z}/p\mathbb{Z})) \times \mathbb{Z}$. Then*

$$S \not\subseteq T \implies F(H_S, 2) \text{ is } \mu_S, F(H_T, 2)\text{–ergodic},$$

where μ_S is the usual product measure on 2^{H_S}.

In particular,

$$S \subseteq T \iff F(H_S, 2) \leq_B F(H_T, 2).$$

Among several other set-theoretic (and measure-theoretic) rigidity results proved in [HK4], we state the following.

Theorem 6.12 ([HK4, Theorem 7]) *Let for $n \geq 1$, $S_n = F(\mathbb{F}_2^n, 2)$ and $R_n = F(\mathbb{F}_2, 2)^n$ (thus $R_1 = S_1$). Then*

$$R_1 <_B R_2 <_B \cdots <_B R_n <_B \cdots, \quad S_1 <_B S_2 <_B \cdots <_B S_n <_B \cdots,$$

and for each n,

$$R_n \leq_B S_n;$$

but for $2 \leq n < m$,

$$S_n \not\leq_B R_m, R_m \not\leq_B S_n.$$

6.5 Torsion-Free Abelian Groups of Finite Rank

We next discuss rigidity results in algebra. The classification problem for torsion-free abelian groups of finite rank is a classical problem in group theory. For rank 1, the problem was solved by Baer in 1937, but no reasonable classification has been found for rank 2 or more; see [T3], [T5] for more on the history of this problem. In the following, denote by $\cong_n$ the isomorphism relation for torsion-free abelian groups of rank $n \geq 1$. As explained in Remark 3.26, we can view this (up to Borel bireducibility) as a countable Borel equivalence relation. Baer's result now implies the following:

Theorem 6.13 (Baer) $\cong_1 \sim_B E_0$.

Using methods from ergodic theory, Hjorth in [H1] showed that $\cong_1 <_B \cong_2$, and finally Thomas proved the following, also using methods from ergodic theory:

Theorem 6.14 ([T3]) *For each $n \geq 1$, $\cong_n <_B \cong_{n+1}$.*

In particular, the Borel cardinality of the set of isomorphism classes of torsion-free abelian groups of rank n encodes the dimension n. Also one can interpret this result as a strong indication of the nonexistence of a reasonable classification for the rank 2 or more case.

For other discussions of these results, see [T5], [Co4], [Q], and [TS].

Denote by $\cong_n^*$ the isomorphism relation on torsion-free abelian groups of rank n that are rigid (i.e., the only automorphisms are the identity and the map $a \mapsto -a$). Then we have:

Theorem 6.15 ([AK]) *For each* $n \geq 1$, $\cong_n^* <_B \cong_{n+1}^*$.

Theorem 6.16 ([T3]) *For each* $n \geq 1$, $\cong_{n+1}^* \not\leq_B \cong_n$.

For each set of primes S, an abelian group is **S-local** if it divisible by any prime $p \notin S$. Denote by $\cong_n^S$ the isomorphism relation on the S-local torsion-free abelian groups of rank n. Also let $\cong_n^p = \cong_n^S$, where $S = \{p\}$. Then we have:

Theorem 6.17 ([T3]) *For each prime p and* $n \geq 1$, $\cong_n^p <_B \cong_{n+1}^p$.

Theorem 6.18 ([T13]) *For sets of primes S, T, and* $n \geq 2$,

$$S \subseteq T \iff \cong_n^S \leq_B \cong_n^T .$$

In [T3, 5.7] it is shown that if E_p is the equivalence relation induced by the action of $\mathrm{GL}_2(\mathbb{Q})$ by fractional linear transformations on $\mathbb{Q}_p \cup \{\infty\}$, then $E_p \sim_B \cong_2^p$, so if $p \neq q$, then E_p, E_q are incomparable in $\leq_B$.

We also have:

Theorem 6.19 ([Co2, Theorem B]) *Let* $n, m \geq 3$, *and* $p \neq q$ *be primes. Then*

$$\cong_m^p \not\leq_B \cong_n^q, \ \cong_n^q \not\leq_B \cong_m^p .$$

For other related results, see [Co1] and [Co3].

6.6 Multiples of an Equivalence Relation

For any countable Borel equivalence relation E and $n \geq 1$, recall that nE is the direct sum of n copies of E. Note that $nE_0 \sim_B E_0$ and $nE_\infty \sim_B E_\infty$. In contrast, we have:

Theorem 6.20 ([T4, 4.9]) *There is a countable Borel equivalence relation E such that*

$$E_0 <_B E <_B 2E <_B 3E <_B \cdots .$$

See also [HK4, 3.9] and [Q, 3.4.10].

7

Hyperfiniteness

7.1 Characterizations and Classification

Recall the following definition:

Definition 7.1 A countable Borel equivalence relation E is **hyperfinite** if $E = \bigcup_n E_n$, where $E_n \subseteq E_{n+1}$ and each E_n is a finite Borel equivalence relation.

We next state a number of equivalent characterizations of hyperfiniteness.

Theorem 7.2 *Let E be a countable Borel equivalence relation on a standard Borel space X. Then the following are equivalent:*

(i) *E is hyperfinite.*
(ii) *$E = \bigcup_{n \geq 1} E_n$, with (E_n) an increasing sequence of finite Borel equivalence relations, such that each E_n-class has cardinality at most n.*
(iii) *$E = \bigcup_n E_n$, with (E_n) an increasing sequence of smooth Borel equivalence relations.*
(iv) *There is a Borel action a of $\mathbb{Z}$ on X such that $E = E_a$.*
(v) *There is a Borel binary relation $R \subseteq X^2$ such that $xRy \implies xEy$ and for each E-class C, $R \restriction C = R \cap C^2$ is a linear ordering on C of order type $\mathbb{Z}$ or finite.*
(vi) *$E \sqsubseteq_B E_0$.*
(vii) *$E \leq_B E_0$.*

In Theorem 7.2, the equivalence of (i) and (ii) is due to Weiss [We1, page 420] (see also [DJK, Theorem 5.1]), (iv) $\implies$ (i) is due to Weiss [We1, Section 4, Theorem 6] and Slaman–Steel [SlSt, Lemma 3.1], and (i) $\implies$ (iv) is due to Slaman–Steel [SlSt, Theorem 3.1]. The equivalence of (i), (iii), (vi), and (vii) is due to Dougherty–Jackson–Kechris [DJK, Theorem 5.1 and Theorem 7.1].

Another proof of the equivalence of (i) and (vi), due to Hjorth, can be found in [Ts1].

We thus have the following complete classification of hyperfinite Borel equivalence relations up to Borel bireducibility.

Corollary 7.3 ([DJK]) *Every hyperfinite Borel equivalence relation is Borel bireducible to exactly one of the equivalence relations in the list*

$$\Delta_1 <_B \Delta_2 <_B \cdots <_B \Delta_n <_B \cdots <_B \Delta_{\mathbb{N}} <_B \Delta_{\mathbb{R}} <_B E_0.$$

Moreover, if E is a nonsmooth hyperfinite Borel equivalence relation, then $E \simeq_B E_0$.

Hyperfinite Borel equivalence relations have also been completely classified up to Borel isomorphism. The main result is the following, where, for any set S, $\text{card}(S)$ is the cardinality of S.

Theorem 7.4 ([DJK, Theorem 5.1]) *Let E, F be aperiodic nonsmooth hyperfinite Borel equivalence relations. Then*

$$E \cong_B F \iff \text{card}(\text{EINV}_E) = \text{card}(\text{EINV}_F).$$

Another version of the proof can be also found in [S15, Section 3.7].

Note that for any countable Borel equivalence relation E,

$$\text{card}(\text{EINV}_E) \in \{0, 1, 2, \ldots, n, \ldots, \aleph_0, 2^{\aleph_0}\}.$$

The tail equivalence relation E_t is hyperfinite, see [DJK, Section 8], and, being compressible, it has no invariant probability measure. Also E_0 has a unique invariant, and thus ergodic, probability measure, nE_0 has exactly n ergodic invariant probability measures, $\mathbb{N}E_0$ has $\aleph_0$ such measures, and $F(\mathbb{Z}, 2)$ has $2^{\aleph_0}$ such measures (consider product measures corresponding to the $(p, 1-p)$ measure on $\{0, 1\}$ for $0 < p < 1$). Another example of a hyperfinite Borel equivalence relation with $2^{\aleph_0}$ ergodic invariant probability measures is $\mathbb{R}E_0$, where for each equivalence relation E, $\mathbb{R}E = E \times \Delta_{\mathbb{R}}$. We thus have:

Corollary 7.5 ([DJK, Corollary 9.3]) *Every aperiodic nonsmooth hyperfinite Borel equivalence relation is Borel isomorphic to exactly one of the equivalence relations in the list*

$$E_t \sqsubset_B^i E_0 \sqsubset_B^i 2E_0 \sqsubset_B^i \cdots \sqsubset_B^i nE_0 \sqsubset_B^i \cdots \sqsubset_B^i \mathbb{N}E_0 \sqsubset_B^i \mathbb{R}E_0 \cong_B F(\mathbb{Z}, 2).$$

In what follows, for each countable Borel equivalence relation E, let EINV_E^0 be the set of E-ergodic E-invariant *nonatomic* probability measures, and put

(i) $c_n(E) = \text{card}(\{C \in X/E\colon \text{card}(C) = n\})$, for $1 \leq n \leq \aleph_0$.
(ii) $s(E) = 0$, if E is smooth; $= 1$, otherwise.

(iii) $t(E) = \text{card}(\text{EINV}^0_E)$.

Corollary 7.6 ([DJK, Corollary 9.2]) *The list* $(c_n)_{1 \leq n \leq \aleph_0}, s, t$ *is a complete list of invariants for Borel isomorphism of hyperfinite Borel equivalence relations.*

A hyperfinite Borel equivalence E is **invariantly universal hyperfinite** if for every hyperfinite Borel equivalence relation F, $F \sqsubseteq^i_B E$. It is easy to see that $E(\mathbb{Z}, \mathbb{R})$ has this property. Clearly an invariantly universal hyperfinite equivalence relation is unique up to Borel isomorphism, and it will be denoted by $E_{\infty h}$. The next corollary gives another manifestation of $E_{\infty h}$.

Corollary 7.7 $E_{\infty h} \cong_B \bigoplus_{n \geq 1} \mathbb{R}\Delta_n \oplus \mathbb{R}E_0$.

The proof of Theorem 7.4 uses, among other tools, the following classical result of Dye in ergodic theory.

Theorem 7.8 ([Dy]) *Let* E, F *be hyperfinite Borel equivalence relations on standard Borel spaces* X, Y, *resp., and let* $\mu \in \text{EINV}_E$, $\nu \in \text{EINV}_F$ *be nonatomic. Then there are an* E*-invariant Borel set* $X_0 \subseteq X$, *an* F*-invariant Borel set* $Y_0 \subseteq Y$, *with* $\mu(X_0) = \nu(Y_0) = 1$, *and a measure-preserving Borel isomorphism* $T \colon E|X_0 \cong_B F|Y_0$.

Equivalently this theorem says that any two Borel $\mathbb{Z}$-actions that have nonatomic ergodic invariant probability measures are orbit equivalent. For a proof, see, e.g., [KM1, Section 7].

An important ingredient in the proof of Theorem 7.8 is the classical Rokhlin's lemma. We state it in what follows in a strong uniform version proved in [GW]. Other references for this are [S15, 3.4.3] and the Corrections and Updates of [KM1, 7.5] that are posted in: `pma.caltech.edu/people/alexander-kechris`.

In the following, a Borel automorphism T of a standard Borel space is called **aperiodic** if all its orbits are infinite.

Theorem 7.9 ([GW, 7.9]) *Suppose* T *is an aperiodic Borel automorphism of a standard Borel space* X, $n \geq 1$, *and* $\epsilon > 0$. *Then there is a Borel complete section* $A \subseteq X$ *of the equivalence relation induced by* T *such that:*

(i) $T^i(A) \cap T^j(A) = \emptyset$, *if* $0 \leq i < j < n$

and, for any T*-invariant probability measure* μ,

(ii) $\mu(X \setminus \bigcup_{i<n} T^i(A)) < \epsilon$.

Suppose E, F are countable Borel equivalence relations on standard Borel spaces X, Y, resp. Let $\mu \in \text{EQINV}_E$, $\nu \in \text{EQINV}_F$. We say that (E, μ) is

isomorphic to (F, ν) if there is an E-invariant Borel set $X_0 \subseteq X$ and an F-invariant Borel set $Y_0 \subseteq Y$ with $\mu(X_0) = \nu(Y_0) = 1$ and $T\colon E|X_0 \cong_B F|Y_0$ such that $T_*\mu \sim \nu$. The **Dye–Krieger Classification** provides a classification up to isomorphism of such (E, μ) for hyperfinite E. One says that (E, μ) is of type $\mathbf{I}_n$, for $1 \leq n \leq \aleph_0$, if μ is atomic concentrating on an E-class of cardinality n. Up to isomorphism there is exactly one (E, μ) of type I_n.

From now on assume that μ is nonatomic. Then (E, μ) is of type $\mathbf{II}_1$ if there is $\nu \sim \mu$ with $\nu \in \mathrm{EINV}_E$, and it is of type $\mathbf{II}_\infty$ if there is an infinite E-invariant measure $\nu \sim \mu$ (i.e., $\nu(X) = \infty$). Otherwise (E, μ) is of type **III**. Dye's Theorem implies that, up to isomorphism, there is exactly one (E, μ) of type II_1 and exactly one (E, μ) of type II_∞. The type-III equivalence relations are further subdivided into classes $\mathbf{III}_\lambda$, for $\lambda \in [0, 1]$. Krieger showed that for $\lambda > 0$, there is a unique, up to isomorphism, (E, μ) of type III_λ, while there is a bijection between isomorphism classes of III_0 equivalence relations and free Borel actions of $\mathbb{R}$ with nonatomic ergodic quasi-invariant probability measure up to isomorphism (of the actions). For a proof of Krieger's results, see [Kr] and [KW].

Another corollary of Theorem 7.4 is the following:

Corollary 7.10 ([DJK, Corollary 9.7]) *Let E be an aperiodic hyperfinite Borel equivalence relation. Then for any sequence (M_n) of positive integers ≥ 2, there is an increasing sequence (E_n) of finite Borel subequivalence relations of E such that $E = \bigcup_n E_n$ and every E_n-class has cardinality $M_0 M_1 \cdots M_n$.*

Compare this with Corollary 2.20 and Theorem 7.2(ii).

A further application of Theorem 7.4 is a classification of nonsmooth Borel equivalence relations induced by Borel actions of $\mathbb{R}$.

Theorem 7.11 ([Ke4, Theorem 3]) *Let a, b be two Borel actions of $\mathbb{R}$ on standard Borel spaces. Let c_a be the cardinality of the set of singleton orbits of the action a and similarly for b. Then if E_a, E_b are not smooth, $E_a \cong_B E_b \iff c_a = c_b$.*

In particular, for any two actions of $\mathbb{R}$ as above with uncountable orbits and E_a, E_b not smooth, $E_a \cong_B E_b$. Using Theorem 3.12 and [JKL, 1.15, 1.16], this holds as well for any such actions of a Polish locally compact group that is compactly generated of polynomial growth.

An analog of Theorem 7.4 for classification of nonsmooth E_a, where a is a free Borel action of $\mathbb{R}^n$, up to Lebesgue orbit equivalence, i.e., Borel isomorphism that preserves the Lebesgue measure on each orbit, is given in [Sl1] and [Sl2]. For results concerning time-change equivalence of free Borel actions of $\mathbb{R}^n$, see [MR2] and [Sl4].

Another analog of Theorem 7.4 is found in [DJK, Section 10], which provides a classification (up to Borel isomorphism) of Lipschitz homeomorphisms of $2^{\mathbb{N}}$.

For countable Borel equivalence relations E, F, we put $E \subseteq_B F$ if there is $E' \subseteq F$ with $E \cong_B E'$. Then $\subseteq_B$ is a quasi-order called the **Borel inclusion order**. Denote by $\subset_B$ its strict part. Using Theorem 7.4, the following was shown in [FKSV].

In what follows, for a quasi-order $\leq$ with strict part $<$ on a set Q and $q, r \in Q$, we say that r is a **successor** to q if $q < r$ and $(q < s \leq r \implies r \leq s)$.

Theorem 7.12 ([FKSV, 2.2.5])

(i) $\mathbb{R}E_0 \subset_B \mathbb{N}E_0 \subset_B \cdots \subset_B 3E_0 \subset_B 2E_0 \subset_B E_0 \subset_B E_t$, *each equivalence relation in this list is a successor in* $\subseteq_B$ *of the one preceding it and* $\mathbb{N}E_0$ *is the infimum in* $\subseteq_B$ *of the* $nE_0, n \in \mathbb{N} \setminus \{0\}$.

(ii) $\mathbb{R}I_{\mathbb{N}} \subset_B E_t$ *and* E_t *is a successor of* $\mathbb{R}I_{\mathbb{N}}$ *in* $\subseteq_B$.

Corollary 7.13 ([FKSV, 2.2.6]) *Let E, F be nonsmooth aperiodic hyperfinite Borel equivalence relations on uncountable standard Borel spaces. Then*

$$E \subseteq_B F \iff \text{card}(\text{EINV}_E) \geq \text{card}(\text{EINV}_F).$$

The next result is a version of the Glimm–Effros dichotomy (see Theorem 5.5) for the inclusion order $\subseteq_B$ instead of $\sqsubseteq_B$.

Corollary 7.14 ([FKSV, 2.3.2]) *Let E be an aperiodic countable Borel equivalence relation. Then exactly one of the following holds:*

(i) *E is smooth.*

(ii) $\mathbb{R}E_0 \subseteq_B E$.

Finally we mention the following characterization of smoothness for aperiodic hyperfinite Borel equivalence relations.

Theorem 7.15 ([KST, 1.1]) *Let E be an aperiodic hyperfinite Borel equivalence relation on a standard Borel space X. Then the following are equivalent:*

(i) *E is smooth.*

(ii) *For every partition of X into Borel sets $X = A \sqcup B$ such that both A, B have infinite intersection with every E-class, we have that $A \sim_E B$.*

7.2 Hyperfinite Subequivalence Relations

Recall from Corollary 2.20 that every aperiodic countable Borel equivalence relation contains an aperiodic hyperfinite Borel subequivalence relation. The following is a strengthening of this result:

Theorem 7.16 *Let E be an aperiodic countable Borel equivalence relation. Then there is an aperiodic hyperfinite Borel equivalence relation $F \subseteq E$ such that* $\mathrm{INV}_E = \mathrm{INV}_F$*, and so* $\mathrm{EINV}_E = \mathrm{EINV}_F$*, and E, F have the same ergodic decomposition (as in Theorem 4.12) modulo compressible sets (for E or equivalently F).*

For a proof, see [Ke5, 5.66 and paragraph following it] or [FKSV, 2.2.3]. The proof is based on the following result related to Theorem 7.8 (and has a similar proof).

Theorem 7.17 (see [Ke5, 5.26]) *Let E be an aperiodic countable Borel equivalence relation on a standard Borel space X, and let $\nu \in \mathrm{EINV}_E$. Then there are an E-invariant Borel set $X_0 \subseteq X$ and an E_0-invariant Borel set $Y_0 \subseteq 2^{\mathbb{N}}$ with $\nu(X_0) = \mu(Y_0) = 1$, where μ is the usual product measure on $2^{\mathbb{N}}$, and a measure preserving Borel isomorphism $T\colon Y_0 \to X_0$ such that $x, y \in Y_0, xE_0y \implies T(x)ET(y)$.*

The fact that every aperiodic countable Borel equivalence relation contains an aperiodic hyperfinite Borel equivalence relation admits the following generalization.

Theorem 7.18 ([CM1, 2.5.1]) *Let E be a countable Borel equivalence relation on a standard Borel space and $\rho\colon E \to \mathbb{R}^+$ be a Borel cocycle for which E is ρ-aperiodic. Then there is a hyperfinite Borel subequivalence relation $F \subseteq E$ for which F is $\rho \restriction F$-aperiodic.*

From Theorem 7.18 the following is also derived:

Theorem 7.19 ([CM1, 2.5.3]) *Let E be a countable Borel equivalence relation on a standard Borel space X, and let $\mu \in \mathrm{QINV}_E$ be such that there is no Borel set of μ-positive measure A for which $E \restriction A$ is smooth. Then there is a hyperfinite Borel subequivalence relation $F \subseteq E$ with the same property.*

Actually in [CM1, 2.5.1, 2.5.3] stronger statements are proved concerning Borel graphs.

7.3 Generic Hyperfiniteness

A Borel equivalence relation is **essentially hyperfinite** if it is Borel bireducible with a hyperfinite Borel equivalence relation, and it is **reducible-to-hyperfinite** if it is Borel reducible to a hyperfinite Borel equivalence relation. In view of Theorems 7.2 and 5.5, a Borel equivalence relation E is essentially hyperfinite if and only if it is reducible-to-hyperfinite.

The following result shows that essential hyperfiniteness always holds generically.

Theorem 7.20 ([HK1, 6.2]) *Let E be a Borel equivalence relation on a Polish space X such that E is reducible-to-countable. Then there is a comeager E-invariant Borel set $C \subseteq X$ such that $E \restriction C$ is essentially hyperfinite. In particular, for any countable Borel equivalence relation E on a Polish space X, there is an E-invariant Borel set $C \subseteq X$ such that $E \restriction C$ is hyperfinite.*

This extends an earlier result of [SWW], and its proof also uses the following result, which is an analog of Proposition 4.17.

Proposition 7.21 (Woodin, see [HK1, 6.5]) *Let E be a countable Borel equivalence relation on a Polish space X.Then there is a comeager complete Borel section $C \subseteq X$ such that for every meager Borel set $A \subseteq C$, $[A]_E$ is meager.*

Another proof of Theorem 7.20 for countable E can be found in [KM1, 12.1].

The second part of the following result was originally proved in [SWW] and the first part by Woodin. A countable Borel equivalence relation E on a Polish space X is called **generically ergodic** if every E-invariant Borel set is meager or comeager. It is called **generic** if the E-saturation of a meager set is meager.

Theorem 7.22 ([SWW]; Woodin, see [Ke5, 5.44–5.46]) *Let E be a a generically ergodic countable Borel equivalence relation on a perfect Polish space X. Then there are a dense G_δ set $X_0 \subseteq X$, an E_0-invariant dense G_δ set $Y_0 \subseteq 2^{\mathbb{N}}$, and a homeomorphism $f \colon X_0 \to Y_0$ which takes $E \restriction X_0$ to $E_0 \restriction Y_0$. If, moreover, E is generic, the set X_0 can be also be taken to be E-invariant.*

We note that Theorem 7.20 fails for measure instead of for category. For example, consider the equivalence relation $E(\mathbb{F}_2, 2)$ on $X = 2^{\mathbb{F}_2}$ with the usual product measure. Then it is a standard result that for any Borel set $A \subseteq X$ of positive measure, $E \restriction A$ is not hyperfinite (see, e.g., Proposition 7.30 following).

In fact Theorem 7.20 fails for measure, even for *compressible* equivalence relations. There is indeed a compressible countable Borel equivalence relation E on a Polish space X and a probability measure μ on X for which there is no invariant Borel set $A \subseteq X$ with $\mu(A) = 1$ and $E|A$ hyperfinite; see [CG, Théorème 2] and [Mo, Section 4].

7.4 Closure Properties

Next we state the basic closure properties of hyperfiniteness:

Theorem 7.23 ([DJK, 5.2], [JKL, 1.3], Theorem 2.37)

(i) *If E, F are countable Borel equivalence relations, F is hyperfinite, and $E \leq_B^w F$, then E is hyperfinite.*
(ii) *If $E \subseteq F$ are countable Borel equivalence relations, E is hyperfinite, and every F-class contains only finitely many E-classes, then F is hyperfinite.*
(iii) *If each E_n, $n \in \mathbb{N}$, is a hyperfinite Borel equivalence relation, then $\bigoplus_n E_n$ is hyperfinite.*
(iv) *If E, F are hyperfinite Borel equivalence relations, then $E \times F$ is hyperfinite.*

The fundamental open problem concerning the closure properties of hyperfiniteness is the following:

Problem 7.24 Let E_n, $n \in \mathbb{N}$, be hyperfinite Borel equivalence relations such that $E_n \subseteq E_{n+1}$ for every n. Is $\bigcup_n E_n$ hyperfinite?

An interesting example of an equivalence relation that is the union of an increasing sequence of hyperfinite Borel equivalence relations was discovered in [Sm]. Fix a countable standard model M of set theory, and let X be the space of Cohen generic reals over M, a G_δ subset of $2^{\mathbb{N}}$. On X define the following equivalence relation:

$$xEy \iff M[x] = M[y].$$

Then it is shown in [Sm] that E is the union of an increasing sequence of hyperfinite Borel equivalence relations. It is not known if E is hyperfinite.

The following notion was introduce in [BJ2]. Here, for $x, y \in \mathbb{N}^{\mathbb{N}}$, we let

$$x \leq_* y \iff \exists m \forall n \geq m (x_n \leq y_n)$$

and recall that $E_0(\mathbb{N})$ is the eventual equality relation on $\mathbb{N}^{\mathbb{N}}$.

Let E be a countable Borel equivalence relation on a standard Borel space X. Then E is **Borel-bounded** if for every Borel function $f: X \to \mathbb{N}^{\mathbb{N}}$, there is a Borel function $g: X \to \mathbb{N}^{\mathbb{N}}$ such that $\forall x(f(x) \leq_* g(x))$ and $g: E \to_B E_0(\mathbb{N})$. Boykin and Jackson [BJ2] show that hyperfinite Borel equivalence relations are Borel bounded and that the closure properties (i), (ii), and (iii) of Theorem 7.23 hold if hyperfiniteness is replaced by Borel-boundedness. However, the analog of part (iv) remains open. They also show that if a Borel-bounded countable Borel equivalence relation is the union of an increasing sequence of hyperfinite Borel equivalence relations, then it is hyperfinite. It is unknown if *every* countable Borel equivalence relation is Borel-bounded. Thomas in [T11, 5.2] has shown that Martin's conjecture on functions on Turing degrees implies that $\equiv_T$ is not Borel-bounded.

We recall here the statement of **Martin's conjecture**. For $x, y \in 2^{\mathbb{N}}$, let $x \leq_T y \iff x$ is recursive in y. Then Martin's conjecture states that for every Borel homomorphism $f\colon \equiv_T \to_B \equiv_T$, either there is $x \in 2^{\mathbb{N}}$ and $z_0 \in 2^{\mathbb{N}}$ such that for all $y \geq_T x$, $f(y) \equiv_T z_0$ or else there is $x \in 2^{\mathbb{N}}$ such that for all $y \geq_T x$, $y \leq_T f(y)$. Equivalently this is stated as follows. Let $\mathcal{D} = 2^{\mathbb{N}}/\equiv_T$ be the set of **Turing degrees** equipped with the partial order $\leq$ induced by $\leq_T$. A **cone** is a subset of the form $\{d \in \mathcal{D} \colon c \leq d\}$, for some $c \in \mathcal{D}$. Then Martin's conjecture says that for every function $f\colon \mathcal{D} \to \mathcal{D}$ that has a Borel lifting, there is a cone C of Turing degrees such that either $f \upharpoonright C$ is constant or else $d \leq f(d)$, for all $d \in C$.

It is also unknown if every Borel-bounded countable Borel equivalence relation is hyperfinite. So we have the following problem:

Problem 7.25 What is the extent of the class of Borel-bounded countable Borel equivalence relations? Are they all hyperfinite, or is every countable Borel equivalence relation Borel-bounded?

For further results on a weakening of the notion of Borel boundedness that holds for all countable Borel equivalence relations and other related notions, see [BJ2].

In [CS] the authors introduce and study certain properties of countable Borel equivalence relations that relate to cardinal characteristics of the continuum.

7.5 μ-Hyperfiniteness

Let E be a countable Borel equivalence relation on a standard Borel space X and μ be a probability measure on X. We say that E is **μ-hyperfinite** if there is an E-invariant Borel set A with $\mu(A) = 1$ such that $E \upharpoonright A$ is hyperfinite. Also E is **measure hyperfinite** if it is μ-hyperfinite for every probability measure μ.

We now state some equivalent conditions for μ-hyperfiniteness.

Proposition 7.26 *Let E be a countable Borel equivalence relation on a standard Borel space X, and let μ be a probability measure on X. Then the following are equivalent:*

(i) *E is μ-hyperfinite.*
(ii) *For any $f_1, f_2, \dots, f_n \in [[E]]_B$, $f_i \colon A_i \to B_i$, and any $\epsilon > 0$, there are $T_1, T_2, \dots, T_n \in [E]_B$ such that the group generated by $T_1, T_2, \dots, T_n$ is finite and for each $1 \leq i \leq n$, $\mu(\{x \in A_i \colon f_i(x) \neq T_i(x)\}) < \epsilon$.*
(iii) *Same as (ii) but with $f_1, f_2, \dots, f_n \in [E]_B$.*
(iv) *Let $T_n \in [E]$ be such that $xEy \iff \exists n(T_n(x) = y)$. Define for any Borel subequivalence relation F of E and $n \geq 1$, $d_n(F, E) = \mu(\{x \in X \colon \exists i < n(\neg xFT_i(x))\})$. Then for each $\epsilon > 0, n \geq 1$, there is a finite Borel subequivalence relation F of E such that $d_n(F, E) < \epsilon$.*

For the proof of Proposition 7.26, see [FM, Section 4], [Ke5, 5.K, A)], [Ts2, Proposition 4], and [M4, Lemma 3.1].

Using Proposition 7.26, one can see that the answer to Problem 7.24 is positive in the measure-theoretic category.

Theorem 7.27 (Dye, Krieger) *Let X be a standard Borel space and μ be a probability measure on X. If $E_n, n \in \mathbb{N}$, is an increasing sequence of μ-hyperfinite Borel equivalence relations, then $\bigcup_n E_n$ is μ-hyperfinite.*

Another proof of this result can be given using the concept of Borel-boundedness. In fact one has the following more general result. In the following, if C is a class of countable Borel equivalence relations, denote by HYP(C) the class of all countable Borel equivalence relations that can be written as the union of an increasing sequence of equivalence relations in C. Then we have:

Theorem 7.28 ([BJ2, page 116]) *Let C be a class of countable Borel equivalence relations closed under subrelations and countable direct sums. Then for any countable Borel equivalence relation E on a standard Borel space X and every probability measure μ on X, if $E = \bigcup_n E_n$, with (E_n) increasing and $E_n \in$ HYP(C), then there are a Borel set $A \subseteq X$ such that $\mu(A) = 1$ and a countable Borel equivalence relation $F \in$ HYP(C) such that $E \upharpoonright A = F \upharpoonright A$.*

Extending Theorem 7.27, Miller [Mi19] showed that if a countable Borel equivalence relation is in the closure of the class of all smooth Borel equivalence relations under countable increasing unions and countable intersections, then it is measure hyperfinite.

We conclude with the following basic open problem.

Problem 7.29 Does measure hyperfiniteness imply hyperfiniteness?

7.6 Groups Generating Hyperfinite Relations

By Theorem 7.2(iv), every Borel action of the group $\mathbb{Z}$ generates a hyperfinite Borel equivalence relation. Which countable groups G have the property that all their Borel actions generate hyperfinite Borel equivalence relations? The following is a well-known fact, see, e.g., [JKL, 2.5 (ii)].

Proposition 7.30 *Let G be a countable group, and let a be a free Borel action of G on a standard Borel space that admits an invariant probability measure. Then if E_a is hyperfinite, G is amenable.*

Since any countable group G admits a free Borel action with invariant probability measure, e.g., its shift action on 2^G, restricted to its free part, with the usual product measure, it follows that every nonamenable group has a free Borel action that generates a nonhyperfinite equivalence relation.

Ornstein and Weiss proved the following:

Theorem 7.31 ([OW]) *Let G be an amenable group and consider a Borel action a of G on a standard Borel space X. Then E_a is measure hyperfinite.*

This motivates the following problem of Weiss [We1].

Problem 7.32 Let G be a countable amenable group. Is it true that every Borel action of G generates a hyperfinite equivalence relation?

Weiss proved a positive answer for the finitely generated abelian groups G. This was extended in [JKL] to all finitely generated nilpotent-by-finite groups, which by the result of Gromov are exactly the finitely generated groups of polynomial growth. In fact we have the following more general result concerning locally compact groups.

Let G be a Polish locally compact group, and let μ_G be a right Haar measure on G. Let d be a positive integer. We say that G is **compactly generated of polynomial growth d** if there is a symmetric compact neighborhood K of the identity of G such that $G = \bigcup_n K^n$ and $\mu_G(K^n) \in O(n^d)$. For $c > 0$, G has the **mild growth property of order c** if there is an increasing sequence (K_n) of compact symmetric neighborhoods of the identity, such that: (a) $K_n^2 \subseteq K_{n+1}$; (b) for infinitely many n, $\mu_G(K_{n+4}) \leq c\mu_G(K_n)$; (c) $\bigcup_n K_n = G$.

It can be shown that if G is compactly generated of polynomial growth, then it has the mild growth property, and if a Polish locally compact group G can be written as a union of an increasing sequence (G_n) of Polish locally compact groups that have the mild growth property of the same order c, then so does G, see [JKL, 1.15]. We now have:

Theorem 7.33 ([JKL, 1.16]) *Let G be a locally compact group with the mild growth property. Then any equivalence relation generated by a Borel action of G on a standard Borel space is essentially hyperfinite.*

In particular, as we mentioned before, any equivalence relation generated by a Borel action of a finitely generated nilpotent-by-finite (i.e., polynomial growth) group is hyperfinite, and the same holds for the group $(\mathbb{Q}^n, +)$.

A significant extension of the result about nilpotent-by-finite groups was recently obtained in [BY, Corollary 1.17]. It is shown there that if a countable Borel equivalence relation admits a locally finite Borel graphing (see Section 9.1) of polynomial growth, then it is hyperfinite.

The next step towards a positive answer to Weiss' problem was taken in [GJ].

Theorem 7.34 ([GJ]) *Let G be any countable abelian group. Then the equivalence relation generated by a Borel action of G on a standard Borel space is hyperfinite.*

Hjorth has raised the following problem: Suppose G is an abelian Polish group and let a be a Borel action of G on a standard Borel space. Is it true that if E is a reducible-to-countable Borel equivalence relation and $E \leq_B E_a$, then E is essentially hyperfinite?

A positive answer has been obtained for non-Archimedean groups.

Theorem 7.35 ([DG, 1.4]) *Let G be an abelian non-Archimedean Polish group and let a be a Borel action of G on a standard Borel space. If E is a reducible-to-countable Borel equivalence relation and $E \leq_B E_a$, then E is essentially hyperfinite.*

Another proof of Theorem 7.35 for $E = E_a$ is given in [Gr1, 1.4].

Corollary 7.36 ([DG, 1.3]) *Let G be an abelian non-Archimedean locally compact Polish group and let a be a Borel action of G on a standard Borel space. Then E_a is essentially hyperfinite.*

More recently a positive answer was obtained in [Cot] for any abelian locally compact Polish group and further extended in [Al] to all countable products of abelian locally compact Polish groups.

However, Allison in [Al] also proved that the answer to Hjorth's problem is in general negative by showing that for every treeable (see Chapter 9) countable Borel equivalence relation E there is a Borel action a of an abelian Polish group such that $E \leq_B E_a$. Finally, Frisch and Shinko strengthened Allison's result to show that this actually holds for *every* countable Borel equivalence relation.

Further extending the methods of [GJ], the following was proved. In the following, a countable group G is **locally nilpotent** if all its finitely generated subgroups are nilpotent. This class of groups properly contains the class of countable nilpotent groups.

Theorem 7.37 ([ScSe]) *Let G be a locally nilpotent countable group. Then the equivalence relation generated by a Borel action of G on a standard Borel space is hyperfinite.*

It is also shown in [ScSe] that the equivalence relation induced by a free and continuous action of a countable locally nilpotent group on a zero-dimensional Polish space continuously embeds into E_0. For other such results concerning continuous embeddings and reductions, see [BJ1], [GJ], and [T9].

It is unknown if Weiss' problem has a positive answer for all solvable groups. However, in a recent development, Conley, Jackson, Marks, Seward, and Tucker-Drob, in [CJMST2], introduce the novel concept of **Borel asymptotic dimension** for a Borel action of a countable group and show that when it is finite, then the induced equivalence relation is hyperfinite. They also show that every free Borel action of a polycyclic group has finite Borel asymptotic dimension and thus the induced equivalence relation is hyperfinite. Combining this with work in [ScSe], the freeness assumption can be dropped and one has the following:

Theorem 7.38 ([CJMST2, Corollary 1.9]) *Let G be a polycyclic group. Then the equivalence relation generated by a Borel action of G on a standard Borel space is hyperfinite.*

In particular, the work in [CJMST2] provides the first examples of finitely generated groups of exponential growth that give a positive answer to Weiss' problem. In fact several other classes of countable groups are shown to give a positive answer to Weiss' problem, at least for free actions. Finally, the methods introduced in [CJMST2] are used in this paper to substantially simplify the proofs of earlier results, including Theorem 7.37.

Changing the point of view, a countable group G is called **hyperfinite generating** if for every aperiodic hyperfinite E there is a Borel action of G that generates E. In [FKSV] equivalent formulations of this property are provided and it is shown that all countable groups with an infinite amenable factor are hyperfinite generating, while no infinite countable group with property (T) has this property.

7.7 Examples

We will discuss here hyperfinite Borel equivalence relations that appear in various contexts.

(1) Let $T \colon X \to X$ be a Borel function on a standard Borel space. Consider the Borel action a of the monoid $S = (\mathbb{N}, +, 0)$ on X given by $1 \cdot x = T(x)$. Then $E_{t,a} = E_T$, and $E_{0,a} = E_{0,T} \subseteq E_T$ is the equivalence relation $x E_{0,T} y \iff \exists n (T^n(x) = T^n(y))$. If T is countable-to-1, E_T and thus $E_{0,T}$ are hyperfinite, see [DJK, Section 8]. In particular, E_t is hyperfinite. In fact it turns out that for an arbitrary Borel function T, the equivalence relations E_T and $E_{0,T}$ are **hypersmooth**, i.e., unions of an increasing sequence of smooth Borel equivalence relations, see [DJK, Section 8].

Let E be a Borel equivalence relation on a standard Borel space X, and let $T \colon E \to_B E$. Put

$$x E_t(E,T) y \iff \exists m \exists n (T^n(x) \; E \; T^m(y))$$

and

$$x E_0(E,T) y \iff \exists n (T^n(x) \; E \; T^n(y)).$$

Clearly $E_0(E,T) \subseteq E_t(E,T)$. The following is an open problem:

Problem 7.39 If E is Borel hypersmooth (resp. Borel hyperfinite) and T is a Borel function (resp., countable-to-1 Borel function), are $E_0(E,T)$, $E_t(E,T)$ hypersmooth (resp. hyperfinite)?

As pointed out in [DJK, Section 8], a positive answer for $E_0(E,T)$ implies a positive answer to Problem 7.24 and gives another proof of Theorem 7.34. In the paper [CFW, Corollary 12], a positive answer is given to Problem 7.39, for the hyperfinite case, in the measure-theoretic context, i.e., $E_t(E,T)$ is measure-hyperfinite.

(2) The Vitali equivalence relation E_v is hyperfinite, see [My]. (This also follows from Theorem 7.2(iii) and the fact that $\mathbb{Q} = \bigcup_{n \geq 1} (\mathbb{Z}/n!)$.) The commensurability relation E_c is hyperfinite. This follows from Theorem 7.34.

(3) Consider the action of $\mathrm{GL}_2(\mathbb{Z})$ on $\mathbb{R} \cup \{\infty\}$ by fractional linear transformations (or equivalently the natural action of $\mathrm{GL}_2(\mathbb{Z})$ on the projective space $\mathrm{PG}(1, \mathbb{R})$). Then the associated equivalence relation is hyperfinite. Similarly consider the action of $\mathrm{GL}_2(\mathbb{Z})$ on the unit circle, where, identifying it with the set of rays $t\vec{x}$ ($t > 0$), for $\vec{x} \in \mathbb{R}^2 \setminus \{0\}$, $A \in \mathrm{GL}_2(\mathbb{Z})$ acts on this ray to give the ray $tA(\vec{x})$ $(t > 0)$. This again generates a hyperfinite Borel equivalence relation. For a proof, see [JKL, 1.4, (C) and page 43].

The action of $\mathrm{GL}_2(\mathbb{Z})$ on the unit circle is also **productively hyperfinite** in the terminology of [CM1, Section 2.1], i.e., its product with *any* Borel action of $\mathrm{GL}_2(\mathbb{Z})$ on a standard Borel space also induces a hyperfinite Borel equivalence relation, see [CM1, 2.1.4]. Using this, it is shown in [CM1, 2.2.3] that the action of $\mathrm{GL}_2(\mathbb{Z})$ on $\mathbb{R}^2$ also generates a hyperfinite Borel equivalence relation.

(4) Consider any countable free group $\mathbb{F}_n$ with a fixed set of free generators. Let $\partial\mathbb{F}_n$ be the **boundary** of $\mathbb{F}_n$, i.e., the set of infinite reduced words (x_i), where each x_i is one of the generators or its inverse, and $x_i x_{i+1} \neq 1$. Then $\mathbb{F}_n$ (viewed as the set of finite reduced words) acts on $\partial\mathbb{F}_n$ by left-concatenation and cancellation. The associated equivalence relation is Borel hyperfinite, see [JKL, 1.4, (E)].

It has been an interesting problem to extend this to the action of any finitely generated hyperbolic group on its boundary. In [A4] a positive answer was obtained for any finitely generated hyperbolic group but in the measure-theoretic category, i.e., the associated equivalence relation is measure hyperfinite. A positive answer in the Borel category has been obtained in [HSS] for any cubulated finitely generated hyperbolic group. Finally, in [MS] the problem was solved in full generality by showing that the answer is positive in the Borel category for any finitely generated hyperbolic group. A new proof of this result is given in [NV], where it is actually shown that this action has finite Borel asymptotic dimension. For other related results, see [Marq], [PS], [O], and [EOSS].

(5) Following up on Remark 3.26 and Theorem 6.13, consider the multiplication action of the multiplicative group of nonzero rationals on the space of subgroups of $(\mathbb{Q}, +)$. The corresponding equivalence relation E is Borel bireducible to $\cong_1$ and thus to E_0. In fact it turns out that E restricted to the subgroups different from $\{0\}, \mathbb{Q}$ is Borel isomorphic to E_t.

Consider now the class of **Butler groups**, which are finite-rank torsion-free abelian groups that can be expressed as finite (not necessarily direct) sums of rank 1 subgroups. In [T7] it is shown that the isomorphism equivalence relation on the class of Butler groups is essentially hyperfinite.

(6) Recall here Example 2.2(5). The isomorphism relation on the space of all subshifts of $k^{\mathbb{Z}}$ is Borel bireducible to E_∞, see [Cl2], However there are various Borel classes of subshifts of $2^{\mathbb{Z}}$ for which the isomorphism relation turns out to be hyperfinite (and not smooth). These include:

(i) The class of nondegenerate rank-1 subshifts, see [GH, Theorem 3.17].
(ii) The class of Toeplitz subshifts with separated holes; see [Kay]. See also [ST] for this result in the measurable context and for raising the question of whether isomorphism in the class of *all* Toeplitz subshifts is hyperfinite.

For each infinite countable group G, one can also consider the isomorphism equivalence relation for subshifts of k^G, $k \geq 2$. This is again a countable Borel equivalence relation, and [GJS2, 9.4.3] shows that for all locally finite G, the isomorphism relation on the space of all subshifts is nonsmooth and hyperfinite. However, if G is not locally finite, isomorphism on the space of free subshifts is Borel bireducible to E_∞, see [GJS2, 9.4.9]. Finally in [ST, 1.2] it is shown that for residually finite, nonamenable G, the isomorphism relation on free, Toeplitz subshifts is not hyperfinite.

(7) In the context of Section 3.6, it is shown in [GK, 8.2] that the isometry relation on proper ultrametric Polish spaces is hyperfinite and not smooth.

(8) For essentially hyperfinite Borel equivalence relations that occur in the context of type spaces in model theory, see [KPS, 3.4, 3.6, 3.7, 3.9, 4.5].

(9) Let $\mathbb{G} = (X, R)$ be a **Borel graph** on a standard Borel space X, i.e., X is the set of vertices, and the set of edges $R \subseteq X^2$ is Borel. We let $E_{\mathbb{G}}$ be the equivalence relation whose equivalence classes are the connected components of $\mathbb{G}$. If $\mathbb{G}$ is **locally countable**, i.e., every vertex has countably many neighbors, then $E_{\mathbb{G}}$ is a countable Borel equivalence relation. In [Mi10] it is shown that if there is a Borel way to choose two ends of the graph in each connected component, then $E_{\mathbb{G}}$ is hyperfinite. Moreover it is shown in [Mi10] that if either there are no ends in each connected component or else there is a Borel way to choose at least three but finitely many ends in each connected component, then $E_{\mathbb{G}}$ is smooth. Finally, in the same paper it is shown that the class of all $E_{\mathbb{G}}$, where $\mathbb{G}$ is **locally finite**, i.e., every vertex has finitely many neighbors and has one end in each connected component, coincides with the class of all aperiodic countable Borel equivalence relations.

(10) Let H be a Polish group and $G \trianglelefteq H$ be a countable normal subgroup. We will say that the quotient group H/G is hyperfinite if the coset equivalence relation of G in H is hyperfinite. It is shown in [FS] that H/G is always hyperfinite. In particular, the outer automorphism group of any countable group is hyperfinite.

(11) The three **Thompson groups** $F \leq T \leq V$ can be defined as follows. By a **T-tree** we mean a finite nonempty subtree S of $2^{<\mathbb{N}}$ such that if $s_1, \dots, s_n$ are the terminal nodes of S, then for all $x \in 2^{\mathbb{N}}$, there exists $i \leq n(s_i \subseteq x)$. For such a tree S, we put $|S| = n$ and order its leaves in lexicographical order $s_1 < \cdots < s_n$. For two such trees S, T with $|S| = |T| = n$, let $f_{S,T}$ be the homeomorphism of $2^{\mathbb{N}}$ defined by $f_{S,T}(s_i x) = t_i x$, for $x \in 2^{\mathbb{N}}, 1 \leq i \leq n$. The collection of all $f_{S,T}$ is a subgroup of the homeomorphism group of $2^{\mathbb{N}}$, the Thompson group F.

Now for each S, T as above and $1 \leq m \leq n$, let $c_m \colon \{1, \dots, n\} \to \{1, \dots, n\}$ be the "cyclic" permutation $c_m(i) = (i+m-1)$, if $(i+m-1) \leq n; = (i+m-1)-n$, otherwise. Then define the homeomorphism $f_{S,T,m}$ of $2^{\mathbb{N}}$ by $f_{S,T,m}(s_i x) = t_{c_m(i)} x$, for $1 \leq i \leq n$. The collection of all $f_{S,T,m}$ is a subgroup of the homeomorphism group of $2^{\mathbb{N}}$, the Thompson group T.

Finally, for each S, T as above and π any permutation of $\{1, \dots, n\}$, define the homeomorphism $f_{S,T,\pi}$ of $2^{\mathbb{N}}$ by $f_{S,T,\pi}(s_i x) = t_{\pi(i)} x$, for $1 \leq i \leq n$. The collection of all $f_{S,T,\pi}$ is a subgroup of the homeomorphism group of $2^{\mathbb{N}}$, the Thompson group V.

Each of these groups acts in the obvious way on $2^{\mathbb{N}}$, and we let E_F, E_T, and E_V be the corresponding equivalence relations. Then it turns out that $E_F|(2^{\mathbb{N}} \setminus \{\bar{0}, \bar{1}\}) = E_t|(2^{\mathbb{N}} \setminus \{\bar{0}, \bar{1}\}), E_T = E_V = E_t$, where $\bar{i}$ is the constant sequence with value i.

7.8 Dichotomies for Essential Hyperfiniteness

Recall here the discussion and notation in Section 3.7. Also recall that a Borel equivalence relation E is hypersmooth if it can be written as the union of an increasing sequence of Borel smooth equivalence relations. Then E_1 is hypersmooth, and for any hypersmooth Borel E, we have that $E \leq_B E_1$, see [KL, 1.3]. Recall also from Theorem 7.2(iii) that the hyperfinite Borel equivalence relations are exactly the hypersmooth countable Borel equivalence relations. We now have the following dichotomy:

Theorem 7.40 ([KL, Theorem 1]) *Let E be a Borel equivalence relation such that $E \leq_B E_1$. Then exactly one of the following holds:*

(i) *E is essentially hyperfinite.*
(ii) $E \sim_B E_1$.

We also have an analogous dichotomy theorem for E_3.

Theorem 7.41 ([HK3, 7.1]) *Let E be a Borel equivalence relation such that $E \leq_B E_3$. Then exactly one of the following holds:*

(i) *E is essentially hyperfinite.*
(ii) $E \sim_B E_3$.

Concerning E_2, the following is an open problem:

Problem 7.42 Let E be a Borel equivalence relation such that $E \leq_B E_2$. Is it true that exactly one of the following holds?

(i) E is essentially hyperfinite.
(ii) $E \sim_B E_2$.

By Theorem 3.29, this holds if "hyperfinite" is replaced by "reducible-to-countable."

7.9 Properties of the Hyperfinite Quotient Space

Let E, F be countable Borel equivalence relations on standard Borel spaces X, Y, resp. In what follows, we say that $f\colon (X/E)^n \to Y/F, n \geq 1$, is Borel if it has a Borel lifting. Similarly, a relation $R \subseteq (X/E)^n$ is Borel if its lifting is Borel. A **Borel isomorphism** of X/E with Y/F is a Borel bijection between X/E and Y/F. It follows from Theorem 2.32 and Corollary 7.3 that if E, F are nonsmooth and hyperfinite, then X/E and Y/F are Borel isomorphic. We can thus refer to $2^{\mathbb{N}}/E_0$ as *the* **hyperfinite quotient space**. We will discuss here some properties of this space.

(1) Borel equivalence relations on the hyperfinite space correspond to Borel equivalence relations containing E_0.

The papers [Mi4] and [Mi17] contain a classification of Borel equivalence relations on the hyperfinite space, all of whose classes have fixed cardinality $n \geq 1$, up to Borel isomorphism. The following is then a corollary of this classification:

Theorem 7.43 ([Mi4], [Mi17, Theorem 5]) *For each $n \geq 1$, there are only finitely many, up to Borel isomorphism, Borel equivalence relations on the hyperfinite quotient space all of whose classes have cardinality n.*

(2) For each set X and $n \geq 1$, let $[X]^n = \{(x_i)_{i<n} : \forall i \neq j (x_i \neq x_j)\}$. Now let E be a Borel equivalence relation on a standard Borel space X. Then for $n \geq 1$, we say that E has the ***n*-Jónsson property** if for all functions $f\colon [X/E]^n \to X/E$, there is a set $A \subseteq X/E$ such that there is a bijection between X/E and A, and

$f([A]^n) \neq X/E$. Also E has the **Jónsson property** if the above holds when $[X/E]^n$ is replaced by $\bigcup_{n\geq 1}[X/E]^n$.

As a special case of more general results proved in [HJ] and [CM], we have the following, where **AD** is the **Axiom of Determinacy**:

Theorem 7.44

(i) ([HJ]) (AD) $\Delta_{\mathbb{R}}$ *has the Jónsson property and* E_0 *has the* 2*-Jónsson property.*

(ii) ([CM]) (AD) E_0 *does not have the* 3*-Jónsson property.*

The n-Jónsson property is related to another property called the Mycielski property. Again let $n \geq 1$. We say that E has the ***n*-Mycielski property** if for every comeager Borel set $C \subseteq X^n$, there is a Borel set $A \subseteq X$ such that $E \sim_B E \upharpoonright A$ and $[A]^n_E = \{(x_i)_{i<n} \in X^n\colon \text{ for all } i \neq j(\neg x_i E x_j\} \subseteq C$. A classical result of Kuratowski and Mycielski (see, e.g., [Ke6, 19.1]) asserts that $\Delta_{\mathbb{R}}$ has the n-Mycielski property for every n. We now have (again as special cases of more general results):

Theorem 7.45

(i) ([HJ]) E_0 *has the* 2*-Mycielski property.*

(ii) ([CM]) E_0 *does not have the* 3*-Mycielski property.*

(3) Finally, in [CJMST1] the authors study, under AD, ultrafilters on the hyperfinite space and show that there is such an ultrafilter lying above, in the Rudin–Keisler order, the Martin ultrafilter on $\mathcal{D} = 2^{\mathbb{N}}/\equiv_T$ (i.e., the ultrafilter generated by the cones), see [CJMST1, 1.8].

7.10 Effectivity of Hyperfiniteness

The following problem, raised in [DJK, Section 5], is still open:

Problem 7.46 Let E be a hyperfinite equivalence relation on $\mathbb{N}^{\mathbb{N}}$ that is Δ^1_1 (effectively Borel). Is there a Δ^1_1 automorphism of $\mathbb{N}^{\mathbb{N}}$ such that $E = E_T$? Equivalently, is it true that $E = \bigcup_n E_n$, where (E_n) is a Δ^1_1 (uniformly in n) increasing sequence of finite equivalence relations?

Miri Segal in her Ph.D. Thesis [Se] showed that the answer is positive in the measure-theoretic context:

Theorem 7.47 ([Se]) *Let μ be a probability measure on $\mathbb{N}^{\mathbb{N}}$, and let E be a μ-hyperfinite equivalence relation on $\mathbb{N}^{\mathbb{N}}$ that is Δ^1_1. Then there are a $\Delta^1_1(\mu)$ E-invariant set A, with $\mu(A) = 1$, and a $\Delta^1_1(\mu)$ increasing sequence (E_n) of finite equivalence relations, such that $E \restriction A = \bigcup_n E_n \restriction A$.*

A proof can be found in [Ts2], [CM1, 1.7.8], and [M4, Remark 4.4].

7.11 Bases for Nonhyperfiniteness

Given a quasi-order $\leq$ on a set A, a **basis** for $\leq$ is a subset $B \subseteq A$ such that for all $a \in A$ there exists $b \in B(b \leq a)$. In this terminology, Theorem 5.5 implies that $\{E_0\}$ is a basis for the quasi-order $\leq_B$ on nonsmooth countable Borel equivalence relations. One can now ask whether there is a "reasonable basis" for $\leq_B$ on nonhyperfinite countable Borel equivalence relations. This is a rather vague question, but one can still formulate some precise problems.

Problem 7.48

(i) Is there a countable basis for the quasi-order of Borel reducibility $\leq_B$ on nonhyperfinite Borel equivalence relations?
(ii) Consider the class $\mathcal{B}$ of all equivalence relations of the form E_a, where a is a free Borel action of $\mathbb{F}_2$ admitting an invariant probability measure. Is $\mathcal{B}$ a basis for the quasi-order of Borel reducibility $\leq_B$ on nonhyperfinite Borel equivalence relations?

Although these questions are still open, recent work in [CM1] shows that there are some severe obstacles to a positive answer (at least for part (i)). In what follows, we say that a countable Borel equivalence relation E on a standard Borel space X is **measure reducible** to a countable Borel equivalence relation F, in symbols $E \leq_M F$, if for every probability measure μ on X, there is a Borel set $A \subseteq X$ with $\mu(A) = 1$ such that $E \restriction A \leq_B F$. Analogously, we define the concept of **measure embeddability**, $E \sqsubseteq_M F$, and their strict counterparts $E <_M F$ and $E \sqsubset_M F$

Consider now the class of all countable Borel equivalence relations that are not measure hyperfinite. This is clearly closed upwards under $\leq_M$. We now have the following result from [CM1], but see also the correction in item 26 under Publications in `http://glimmeffros.github.io`:

Theorem 7.49 ([CM1]) *Any basis for the quasi-order of measure reducibility on countable Borel equivalence relations that are not measure hyperfinite has the cardinality of the continuum.*

In particular, this shows that every basis of cardinality less than the continuum for the quasi-order of Borel reducibility $\leq_B$ on nonhyperfinite countable Borel equivalence relations must contain equivalence relations that are measure hyperfinite. It is an open problem whether such relations exist, see Problem 7.29. Also it follows from Theorem 7.49 that in Problem 7.48(ii), one cannot replace $\mathcal{B}$ by a subset that has cardinality less than that of the continuum.

One can also ask the analog of Problem 7.48 for the quasi-order of weak Borel reducibility $\leq_B^w$.

Finally there are basis questions concerning inclusion of equivalence relations instead of reducibility. A version of Problem 7.48(ii) was considered in [KM1, 28.7].

Problem 7.50 Consider the class C of countable Borel equivalence relations E that are not μ-hyperfinite for every E-*invariant* probability measure μ, and let C' be the subclass consisting of all equivalence relations of the form E_a, where a is a free Borel action of $\mathbb{F}_2$. Is C' a basis for C for the partial order $\subseteq$ of inclusion?

For a partial answer, see [KM1, 28.8].

The following is also a related well-known problem in the measure-theoretic context (see, e.g., [KM1, 28.14]). It is a dynamic version of the von Neumann–Day problem that asked whether every nonamenable countable group contains a copy of $\mathbb{F}_2$ (the answer is negative as proved by Ol'shanskii).

Problem 7.51 Let E be a countable Borel equivalence relation on a standard Borel space X, and let μ be an E-invariant, E-ergodic probability measure on X. Is it true that exactly one of the following holds:

(i) E is μ-hyperfinite.
(ii) There are an E-invariant Borel set $A \subseteq X$ with $\mu(A) = 1$ and a free Borel action a of $\mathbb{F}_2$ on A such that $E_a \subseteq E \upharpoonright A$?

It was shown in [GL] that for every countable nonamenable group G, there are *some* standard Borel space X and probability measure ν on X, such that if E is induced by the shift action of G on X^G and $\mu = \nu^G$ is the product measure, then (ii) holds, with μ being also E_a-ergodic. It was shown in [Bo] that this holds for *every* (X, ν), when ν does not concentrate on one point.

Moreover, it is shown in [BHI] that if E is a countable Borel equivalence relation on a standard Borel space X and μ is an E-invariant, E-ergodic probability measure on X such that E is not μ-hyperfinite, then there are a countable Borel equivalence relation F on a standard Borel space Y, an F-invariant, F-ergodic measure ν on Y, a Borel map $f\colon Y \to X$ such that $f_*\nu = \mu$, and a Borel

F-invariant set $A \subseteq Y$ with $\nu(A) = 1$ such that for every $y \in A$, $f \upharpoonright [y]_F$ is a bijection with $[f(y)]_E$ and there is a free Borel action a of $\mathbb{F}_2$ on A such that $E_a \subseteq F \upharpoonright A$, with ν being also E_a-ergodic. More succinctly, this states that E has a class-bijective extension in which Problem 7.51(ii) holds.

7.12 Hyper-Borel-finiteness

The following notion is introduced in [DaMa], which traces its origins to the paper [SlSt]. Let X be a standard Borel space and (f_n) be a sequence of Borel functions $f_n \colon X \to X^{\mathbb{N}}$. A countable Borel equivalence relation E on X is called **(f_n)-finite** if for every $x \in X$ and every n, $f_n(x)(\mathbb{N})$ is not an infinite subset of $[x]_E$. Also E is called **hyper-(f_n)-finite** if E is the union of an increasing sequence of (f_n)-finite subequivalence relations. Finally, E is called **hyper-Borel-finite** if it is hyper-(f_n)-finite for every sequence of Borel functions (f_n). Every hyperfinite equivalence relation is hyper-Borel-finite, but it is unknown whether the converse holds or in fact whether *every* countable Borel equivalence relation is hyper-Borel-finite. It is also unknown how this notion is related to the concept of Borel boundedness discussed in Section 7.4.

In [SlSt, Question 6], the question was raised of whether Turing equivalence $\equiv_T$ on $2^{\mathbb{N}}$ is hyper-(f_n)-finite, where (f_n) is the sequence enumerating the Turing reductions. It is shown in [DaMa, 3.5] that a positive answer is equivalent to the statement that every countable Borel equivalence relation is hyper-Borel-finite.

8

Amenability

8.1 Amenable Relations

Let G be a Polish locally compact group, and let λ be a left Haar measure. Recall that G is **amenable** if there is a finitely additive probability measure defined on all λ-measurable subsets of G, vanishing on λ-null sets, that is invariant under left-translation. We say that a Polish locally compact group G satisfies the **Reiter condition** if there is a sequence (F_n) of Borel functions $F_n : G \to \mathbb{R}$ such that $F_n \geq 0$, $\|F_n\|_1 = 1$ and for all $g \in G(\|F_n - g \cdot F_n\|_1 \to 0)$, where for a function $F : G \to X$, X any set, $g \cdot F : G \to X$ and $g \cdot F(h) = F(g^{-1}h)$.

One of the many equivalent characterizations of amenability is the following (see [Pa, 0.8, Problem 4.1]):

Theorem 8.1 *Let G be a Polish locally compact group. Then the following are equivalent:*

(i) *G is amenable.*
(ii) *G satisfies the Reiter condition.*

We can now use an analog of the Reiter condition to define a notion of amenability for countable Borel equivalence relations (see [Kai] for such a definition in the measure-theoretic context and [JKL, Section 2.4] and [KM1, Section 9]):

Definition 8.2 Let E be a countable Borel equivalence relation. Then E is **amenable** if there is a sequence of Borel functions (f_n) with $f_n : E \to \mathbb{R}$, $f_n \geq 0$ such that letting $f^n_x(y) = f_n(x, y)$, we have for all $x(f^n_x \in \ell^1([x]_E)$, $\|f^x_n\|_1 = 1)$ and $xEy \implies \|f^n_x - f^n_y\|_1 \to 0$.

The following is now immediate, see, e.g., [JKL, 2.13]:

Proposition 8.3 *Let G be a countable amenable group, and let a be a Borel action of G on a standard Borel space. Then E_a is amenable. In particular any hyperfinite Borel equivalence relation is amenable.*

Conversely we have the following, extending Proposition 7.30 (see, e.g., [JKL, 2.14]):

Proposition 8.4 *Let G be a countable group, and let a be a free Borel action of G on a standard Borel space that admits an invariant probability measure. Then if E_a is amenable, G is amenable.*

The following strengthening of Problem 7.32 is also open:

Problem 8.5 Let E be an amenable countable Borel equivalence relation. Is it true that E is hyperfinite?

Proposition 8.3 can be generalized as follows for actions of Polish locally compact groups. In what follows, a Borel equivalence relation is called **essentially amenable** (resp., **reducible-to-amenable**) if it is Borel bireducible (resp., reducible) to an amenable countable Borel equivalence relation.

Proposition 8.6 *Let G be an amenable Polish locally compact group, and let a be a Borel action of G on a standard Borel space X. If $S \subseteq X$ is a countable complete Borel section of E_a, then $E_a|S$ is amenable and thus E_a is essentially amenable.*

Proof Fix a Borel surjective function $\pi : X \to S$ such that $\pi(x)E_a x$, for all $x \in X$. Let (F_n) be a sequence as in the definition of the Reiter condition for G. Let $E = E_a|S$, and for xEy, put $A_{xy} = \{g \in G : g^{-1} \cdot x \in \pi^{-1}(\{y\})\}$. Then define $f_n : E \to \mathbb{R}$ by

$$f_n(x, y) = \int_{A_{xy}} F_n(g)\, \mathrm{d}\lambda(g).$$

Since for each $x \in S$, $\{A_{xy} : xEy\}$ is a Borel partition of G, it is clear that $f_n \geq 0$ and $f^n_x \in \ell^1([x]_E)$, $\|f^n_x\|_1 = 1$. Also given xEy, let $g \in G$ be such that $g \cdot x = y$. Then $\|f^n_x - f^n_y\|_1 \leq \|F_n - g \cdot F_n\|_1 \to 0$. □

Again we have the following generalization of Problem 7.32:

Problem 8.7 Let G be an amenable Polish locally compact group. Is it true that any Borel action of G generates an essentially hyperfinite equivalence relation?

As an application of Proposition 8.6, one can give a stronger version of a result proved in [HK2, 5.C]. Let E be the Borel equivalence relation of isomorphism (conformal equivalence) of domains of the form $\mathbb{C} \setminus S$, for S a discrete subset of $\mathbb{C}$. As explained in [HK2, 5.C], this is Borel isomorphic to the equivalence relation induced by the action of the "$az + b$" group (where $a \in \mathbb{C}, a \neq 0, b \in \mathbb{C}$) on the standard Borel space of discrete subsets of $\mathbb{C}$. Since this group is amenable, it follows that this isomorphism relation is essentially amenable. In [HK2, 5.C] it is shown that it is not smooth and is conjectured to be actually essentially hyperfinite.

Another interesting example of an amenable equivalence relation is in [EH, Theorem 1.2, (i)]. Let G be a countable group with a finite set of generators S, and consider the action of G on the space of ends of this group (for this set of generators) and also its associated action on the space of two-element subsets of the space of ends. Then if there are infinitely many ends, the equivalence relation generated by the action on the space of two-element sets is amenable.

8.2 Fréchet Amenability

We will now discuss a (possibly) wider notion of amenability, introduced in [JKL, Section 2.4].

Recall that a **free filter** on $\mathbb{N}$ is a filter containing the **Fréchet filter**

$$Fr = \{A \subseteq \mathbb{N} : A \text{ is cofinite}\}.$$

Definition 8.8 Let E be a countable Borel equivalence relation on a standard Borel space X. Let $\mathcal{F}$ be a free filter on $\mathbb{N}$. We say that E is **$\mathcal{F}$-amenable** if there is a sequence (f_n) of Borel functions $f_n : E \to \mathbb{R}$, $f_n \geq 0$ such that letting $f_x^n(y) = f_n(x, y)$, we have: for all $x(f_n^x \in \ell_1([x]_E), \|f_n^x\|_1 = 1)$ and $xEy \Rightarrow \|f_x^n - f_y^n\|_1 \to_{\mathcal{F}} 0$.

As usual if $x_n, x \in \mathbb{R}$, then $x_n \to_{\mathcal{F}} x$ means that for every nbhd U of x there is $A \in \mathcal{F}$ such that $n \in A \Rightarrow x_n \in U$. Clearly $x_n \to_{Fr} x$ if and only if $x_n \to x$.

Define a quasi-order between filters on $\mathbb{N}$ by

$$\mathcal{F} \leq \mathcal{G} \iff \exists f : \mathbb{N} \to \mathbb{N}(f^{-1}(\mathcal{F}) \subseteq \mathcal{G})$$

and the corresponding equivalence relation

$$\mathcal{F} \equiv \mathcal{G} \iff \mathcal{F} \leq \mathcal{G} \text{ and } \mathcal{G} \leq \mathcal{F}.$$

Then if E is $\mathcal{F}$-amenable and $\mathcal{F} \leq \mathcal{G}$, E is $\mathcal{G}$-amenable, so $\mathcal{F}$-amenability only depends on the $\equiv$-equivalence class of $\mathcal{F}$.

Next define a transfinite iteration of the Fréchet filter. For two filters $\mathcal{F}, \mathcal{G}$

on $\mathbb{N}$, define their (Fubini) **product** by

$$\mathcal{F} \otimes \mathcal{G} = \{A \subseteq \mathbb{N} : \{m : \{n : \langle m, n \rangle \in A\} \in \mathcal{G}\} \in \mathcal{F}\},$$

where $\langle m, n \rangle$ is a fixed bijection of $\mathbb{N}^2$ with $\mathbb{N}$. We also define for each sequence $(\mathcal{F}_n)$ of filters, the filter

$$\mathcal{F} \otimes (\mathcal{F}_n) = \{A \subseteq \mathbb{N} : \{m : \{n : \langle m, n \rangle \in A\} \in \mathcal{F}_m\} \in \mathcal{F}\}.$$

For each countable limit ordinal λ, fix an increasing sequence $\alpha_0 < \alpha_1 < \cdots < \lambda$ with limit λ, and inductively define the αth **iterated Fréchet filter** Fr_α as follows:

$$\begin{aligned} Fr_1 &= Fr, \\ Fr_{\alpha+1} &= Fr \otimes Fr_\alpha, \\ Fr_\lambda &= Fr \otimes (Fr_{\alpha_n}). \end{aligned}$$

This definition depends on the choice of $\langle m, n \rangle$ and the sequences (α_n), but it can be shown that it is independent up to $\equiv$.

Definition 8.9 Let E be a countable Borel equivalence relation and $1 \leq \alpha < \omega_1$ be a countable ordinal. We say that E is **α-amenable** if E is Fr_α-amenable. It is **Fréchet-amenable** if it is α-amenable for some $1 \leq \alpha < \omega_1$.

Therefore, for any countable Borel equivalence relation E:

$$E \text{ is amenable if and only if } E \text{ is 1-amenable.}$$

By a simple induction on β,

$$\alpha \leq \beta \Rightarrow Fr_\alpha \leq Fr_\beta,$$

and so

$$\alpha \leq \beta \text{ and } E \text{ is } \alpha\text{-amenable} \implies E \text{ is } \beta\text{-amenable.}$$

We also have the analog of Proposition 8.4:

Proposition 8.10 ([JKL, 2.14]) *Let G be a countable group, and let a be a free Borel action of G on a standard Borel space that admits an invariant probability measure. Then if E_a is Fréchet-amenable, G is amenable.*

We now have the following closure properties;

Proposition 8.11 ([JKL, 2.15] for (i)–(vi)) *Let E, F, E_n be countable Borel equivalence relations and $1 \leq \alpha < \omega_1$. Then we have:*

(i) *If F is α-amenable and $E \leq_B^w F$, then E is α-amenable.*

(ii) *If $E \subseteq F$, E is α-amenable and every F-equivalence class contains only finitely many E-classes then F is α-amenable.*
(iii) *If each E_n is α-amenable, so is $\bigoplus_n E_n$.*
(iv) *If E, F are α-amenable, so is $E \times F$.*
(v) *If (E_n) is an increasing sequence, and for each n, E_n is α_n-amenable for some $\alpha_n < \alpha$, then $E = \bigcup_n E_n$ is α-amenable.*
(vi) *If E is α-amenable and $T : E \leq_B^w E$, then $E_t(E,T)$ is $(\alpha+1)$-amenable.*
(vii) *If E is α-amenable and a is a Borel action of an amenable countable group by automorphisms of E, then the expansion $E(a)$ is $(\alpha+1)$-amenable.*

In particular, the union of an increasing sequence of hyperfinite Borel equivalence relations is 2-amenable. It is not known if the union of an increasing sequence of α-amenable Borel equivalence relations is α-amenable.

The following are the two basic problems concerning Fréchet amenability, representing two opposite possibilities.

Problem 8.12 Is Fréchet amenability equivalent to amenability? Moreover, is Fréchet amenability equivalent to hyperfiniteness?

Problem 8.13 Is the transfinite hierarchy of Fréchet amenability proper, i.e., does $\alpha < \beta$ imply that there is a β-amenable Borel equivalence relation that is not α-amenable?

An α-amenable Borel equivalence relation E is **invariantly universal α-amenable** if for every α-amenable Borel equivalence relation F, $F \sqsubseteq_B^i E$. As a special case of a general result, see [CK, Corollary 4.4], such a universal equivalence relation exists and it is of course unique up to Borel isomorphism. It will be denoted by $E_{\infty\alpha}$.

8.3 Amenable Classes of Structures

An important method used to generate Fréchet-amenable countable Borel equivalence relations proceeds through assigning to each equivalence class, in a uniform Borel way, a structure (in the sense of model theory) with special properties. The more detailed study of the structurability of countable Borel equivalence relation will be undertaken in Chapter 13.

Let $L = \{R_i : i \in I\}$ be a countable relational language, where R_i has arity n_i. Let $\mathcal{K}$ be a class of countable structures in L closed under isomorphism. Let E be a countable Borel equivalence relation on a standard Borel space X. An ***L*-structure on *E*** is a Borel structure $\mathbb{A} = \langle X, R_i^{\mathbb{A}} \rangle_{i \in I}$ of L with universe X, i.e., $R_i^{\mathbb{A}} \subseteq X^{n_i}$ is Borel for each $i \in I$, such that $R_i^{\mathbb{A}}(x_1, x_2, \dots, x_{n_i}) \implies$

$x_1 E x_2 E \cdots E x_{n_i}$. Then for each E-class C, we have that $\mathbb{A} \restriction C = \langle C, R_i^{\mathbb{A}} \cap C^{n_i} \rangle_{i \in I}$ is an L-structure with universe C. If now $\mathbb{A} \restriction C \in \mathcal{K}$, for each E-class C, we say that $\mathbb{A}$ is a **$\mathcal{K}$-structure** on E. If E admits such a $\mathcal{K}$-structure, we say that E is **$\mathcal{K}$-structurable**.

For each (nonempty) countable set X, we denote by $\mathrm{Mod}_X(L)$ the space of L-structures with universe X. It can be identified with $\prod_i 2^{X^{n_i}}$, so it is a compact metrizable space. A class of countable structures $\mathcal{K}$ as above is **Borel** (resp., **analytic, coanalytic**) if for each countable X, $\mathcal{K} \cap \mathrm{Mod}_X(L)$ is Borel (resp., analytic, coanalytic) in $\mathrm{Mod}_X(L)$.

We now consider amenability for classes of structures.

Definition 8.14 ([JKL, Section 2.5])

(i) An analytic class of L-structures $\mathcal{K}$ is **α-amenable**, where $\alpha \geq 1$ is a countable ordinal, if for each countable set X, there is a family of maps

$$(f_n^{\mathbb{A}})_{n \in \mathbb{N},\ \mathbb{A} \in \mathcal{K} \cap \mathrm{Mod}_X(L)},$$

such that:

(a) $f_n^{\mathbb{A}} : X \to \mathbb{R}$, $f_n^{\mathbb{A}} \geq 0$, $f_n^{\mathbb{A}} \in \ell^1(X)$ and $\|f_n^{\mathbb{A}}\|_1 = 1$.
(b) The map $f_n \colon (\mathcal{K} \cap \mathrm{Mod}_X(L)) \times X \to \mathbb{R}$ defined by $f_n(\mathbb{A}, x) = f_n^{\mathbb{A}}(x)$ is Borel.
(c) If $\pi : \mathbb{A} \to \mathbb{B}$ is an isomorphism between $\mathbb{A}$ and $\mathbb{B}$, then

$$\|f_n^{\mathbb{A}} - f_n^{\mathbb{B}} \circ \pi\|_1 \to_{Fr_\alpha} 0.$$

(ii) An arbitrary class $\mathcal{K}$ of L-structures is **Fréchet-amenable** if for any analytic class $\mathcal{K}' \subseteq \mathcal{K}$ there is a countable ordinal α (which may depend on $\mathcal{K}'$) such that $\mathcal{K}'$ is α-amenable. (So, in particular, if $\mathcal{K}$ is analytic, it is Fréchet-amenable if and only if it is α-amenable for some α.)

(iii) A countable L-structure $\mathbb{A}$, with universe A, is **α-amenable** if there is a sequence of maps $f_n : A \to \mathbb{R}$ such that $f_n \geq 0$, $f_n \in \ell^1(A)$, $\|f_n\|_1 = 1$, and for every $\pi \in \mathrm{Aut}(\mathbb{A})$, $\|f_n - f_n \circ \pi\|_1 \to_{Fr_\alpha} 0$, where $\mathrm{Aut}(\mathbb{A})$ is the group of automorphisms of $\mathbb{A}$. (This is equivalent to saying that the isomorphism class of $\mathbb{A}$ is α-amenable.)

Again in this definition, "1-amenable" will from now on be called simply "amenable."

We now have:

Proposition 8.15 ([JKL, 2.18]) *Let E be a countable Borel equivalence relation. If E is $\mathcal{K}$-structurable and $\mathcal{K}$ is Fréchet-amenable, then E is Fréchet-amenable. If, moreover, $\mathcal{K}$ is analytic and α-amenable, then E is α-amenable.*

We proceed to describe various Fréchet-amenable classes of countable structures.

Recall that a linear order is **scattered** if it contains no copy of the rational order. The class of scattered linear orders is coanalytic but not Borel. We now have:

Theorem 8.16 ([Ke1], see also [JKL, 2.19]) *The class of countable scattered linear orders is Fréchet-amenable.*

This result was used in [Ke1] to show (assuming that sharps of reals exist) that if one assigns in a Borel way (in the sense described above) a linear order to each Turing degree, then on a cone of Turing degrees this linear order contains a copy of the rationals. Also if E is an aperiodic countable Borel equivalence relation on a Polish space, which is not Fréchet-amenable and admits an ergodic invariant probability measure μ, for example $F(\mathbb{F}_2, 2)$, then, by Theorem 7.20, on a comeager invariant Borel set one can assign in a Borel way to each E-class a copy of the integer order, but for any Borel assignment of a linear order to each E-class, there will be a μ-conull invariant Borel set on which this order contains a copy of the rationals.

Recall that a **tree** $\mathbb{T} = \langle T, R \rangle$ is a connected (undirected) graph with no cycles. Here T is the set of vertices and R is the edge relation. For the concept of the **branching number** of a locally finite $\mathbb{T}$, due to Lyons, see [JKL, 2.20] and references therein. Every locally finite tree of subexponential growth has branching number 1. We now have the following result, which was proved in the measure-theoretic context in [AL].

Theorem 8.17 ([AL], see also [JKL, 2.21]) *The class of infinite locally finite trees of branching number* 1 *is amenable.*

The same result also holds for the class of locally finite connected graphs of strongly subexponential growth, see [JKL, 2.22].

One can completely characterize when an infinite locally finite tree $\mathbb{T}$ is Fréchet-amenable.

Theorem 8.18 ([AL], [Ne], see also [JKL, 2.24]) *Let* $\mathbb{T}$ *be an infinite locally finite tree. Then the following are equivalent:*

(i) $\mathbb{T}$ *is Fréchet-amenable.*
(ii) $\mathbb{T}$ *is amenable.*
(iii) $\mathrm{Aut}(\mathbb{T})$ *(which is a locally compact Polish group) is amenable.*
(iv) *One of the following is invariant under* $\mathrm{Aut}(\mathbb{T})$*:*

  (a) *a vertex;*

(b) *the set of two vertices connected by an edge;*
(c) *an end;*
(d) *a line.*

Finally, we mention the following model-theoretic property of Fréchet-amenable structures. A countable structure $\mathbb{A} = \langle A, \dots \rangle$ has **trivial definable closure** if for every finite $F \subseteq A$ and formula $\phi(x)$ in $L_{\omega_1\omega}$ with parameters in F, if there is a unique $a \in A$ such that $\mathbb{A} \models \phi(a)$, then $a \in F$.

Theorem 8.19 ([CK, 8.18]) *Let $\mathbb{A}$ be an infinite amenable countable structure. Then $\mathbb{A}$ does not have trivial definable closure.*

8.4 The Connes–Feldman–Weiss Theorem

Let E be a countable Borel equivalence relation on a standard Borel space X and μ be a probability measure on X.

We say that E is **μ-amenable** if there is an E-invariant Borel set $A \subseteq X$ with $\mu(A) = 1$ such that $E|A$ is amenable. Similarly we define what it means to say that E is **μ-Fréchet-amenable**.

We will now formulate a number of conditions that turn out to be equivalent to μ-amenability. Then we state the Connes–Feldman–Weiss Theorem, which identifies these conditions with μ-hyperfiniteness.

(1) First we discuss the following condition, due to Zimmer (see, [Z1, 3.1]), which was the original definition of the concept of μ-amenability. It is motivated by the formulation of amenability for countable groups in terms of a fixed point property of affine actions of the group; see [Z2, 4.1.4].

Let B be a separable Banach space, with $\mathrm{LI}(B)$ the group of its linear isometries, which is Polish under the strong operator topology. Let B_1^* be the closed unit ball of the dual B^*, with the weak*-topology. For $T \in \mathrm{LI}(B)$, denote by T^* the adjoint operator restricted to B_1^*, so that T^* is a homeomorphism of B_1^*. If $\alpha : E \to \mathrm{LI}(B)$ is a Borel cocycle, its adjoint cocycle α^* (into the homeomorphism group $H(B_1^*)$ of B_1^*) is defined by $\alpha^*(x, y) = (\alpha(x, y)^{-1})^*$.

For each compact metrizable space C, let $K(C)$ be the compact metrizable space of closed subsets of C. A Borel map $x \mapsto K_x$ from X into $K(B_1^*)$ is a **Borel field** if for all x, K_x is convex and nonempty. A Borel map $S : X \to B_1^*$ is a **section** of K_x if $S(x) \in K_x$, μ-a.e. (x). A Borel field $(K_x)_{x\in X}$ is **α-invariant** if there is an E-invariant Borel set $A \subseteq X$ with $\mu(A) = 1$ such that $x, y \in A, xEy \implies \alpha^*(x, y)(K_x) = K_y$, and a section S is **α-invariant** if there is an E-invariant Borel set $A \subseteq X$ with $\mu(A) = 1$ such that $x, y \in A, xEy \implies \alpha^*(x,y)(S(x)) = S(y)$.

We say that E is **μ-Z-amenable** if for every separable Banach space B and every Borel cocycle $\alpha : E \to \mathrm{LI}(B)$, every α-invariant Borel field $(K_x)_{x \in X}$ has an α-invariant section.

(2) The next condition comes from [CFW]. We say that E is **μ-CFW-amenable** if there is a positive linear operator

$$P \colon L^\infty(E, M_l) \to L^\infty(X, \mu)$$

that sends the constant 1 function to the constant 1 function, such that for every Borel map $f \colon A \to B$ in $[[E]]_B$, we have $P(F^f) = P(F)^f$, where for $F \in L^\infty(E, M_l)$, $F^f(x, y) = F(f^{-1}(x), y)$, if $x \in B$; 0, otherwise, while for $F \in L^\infty(X, \mu)$, $F^f(x) = F(f^{-1}(x))$, if $x \in B$; 0, otherwise.

Remark 8.20 In [CFW] another condition is also considered, which postulates that there is an assignment of means $[x]_E \mapsto \varphi_{[x]_E}$, which is μ-measurable (in the weak sense), i.e., for each bounded Borel map $f \colon E \to \mathbb{C}$, the function $x \mapsto \varphi_{[x]_E}(f_x)$ is μ-measurable. Compare this with the paragraph preceding Theorem 2.13. We will discuss this further in Remark 8.23.

(3) A further condition, originating in [Kai] and [El], is motivated by the following equivalent formulation of amenability for finitely generated groups (the **Følner condition**): Let G be a finitely generated group with finite symmetric set of generators S, and let $\mathrm{Cay}(G, S)$ be the **Cayley graph** of (G, S). For finite $A \subseteq G$, let $\partial(A) = \{g \in G \colon g \in (G \setminus A)$ & there exists $s \in S$ there exists $h \in A(hs = g)\}$ be the boundary of A in the Cayley graph. The **isoperimetric constant** of the Cayley graph is the infimum of the ratios $\frac{|\partial(A)|}{|A|}$ over all finite nonempty subsets A of G. Then G is amenable if and only if the isoperimetric constant of the Cayley graph is 0.

Analogously, for each locally finite Borel graph $\mathbb{G} = \langle X, R \rangle$ with vertex set X and edge set R, we define for each Borel set $A \subseteq X$, the boundary of A in $\mathbb{G}$ by $\partial_{\mathbb{G}}(A) = \{x \in X \colon x \in (X \setminus A)$ & there exists $y \in A(xRy)\}$. Then the **isoperimetric constant** of $\mathbb{G}$ is the infimum of the ratios $\frac{\mu(\partial_{\mathbb{G}}(A))}{\mu(A)}$ over all Borel subsets $A \subseteq X$ of positive measure such that the induced subgraph $\mathbb{G} \restriction A = (A, R \cap A^2)$ has finite connected components.

We say that E is **μ-KE-amenable** if for some locally finite Borel graph $\mathbb{G} = \langle X, R \rangle$ with $E_{\mathbb{G}} = E$ (see also Remark 9.4 here) and any Borel set Y of positive measure, the isoperimetric constant of $\mathbb{G} \restriction Y = \langle Y, R \cap Y^2 \rangle$ (with the normalized probability measure $(\mu \restriction Y)/\mu(Y)$) is 0.

(4) The final condition originates in another characterization of the notion of amenability for countable groups: A countable group G is amenable if and only if every continuous action of G on a compact metrizable space admits an invariant probability measure.

The following condition, due to Furstenberg, see [H6], is as follows. We say that E is **μ-F-amenable** if for any Borel cocycle $\alpha\colon E \to H(K)$, where K is a compact metrizable space and $H(K)$ its homeomorphism group, there is a Borel map $x \mapsto \mu_x$ from X to $P(K)$ such that for some E-invariant Borel set $A \subseteq X$ with $\mu(A) = 1$, we have

$$x, y \in A, xEy \implies \mu_y = \alpha(x, y)_* \mu_x.$$

We now have the following main theorem concerning all these notions:

Theorem 8.21 *Let E be a countable Borel equivalence relation on the standard Borel space X, and let μ be a probability measure on X. Then the following are equivalent:*

(i) *E is μ-hyperfinite.*
(ii) *E is μ-amenable.*
(iii) *E is μ-Fréchet-amenable.*
(iv) *E is μ-Z-amenable.*
(v) *E is μ-*CFW*-amenable.*
(vi) *E is μ-*KE*-amenable.*
(vii) *E is μ-*F*-amenable.*

The equivalence of (i), (ii), (iv), and (v) is due to [CFW]; see also [Kai] for (ii). The equivalence of (i) and (iii) follows from this and the proof of [JKL, 2.13, (ii)]. A proof of the equivalence of (i) and (vi), following up on [Kai] and [El], is contained in [CGMT, Theorem 1.1], where it is also shown that in the definition of μ-KE-amenability one can equivalently replace "for some locally finite Borel graph" by "for all locally finite Borel graphs," and finally the equivalence of these conditions with (vii) is proved in [H6].

In [M4] there is a simpler proof of the equivalence of (i) and (ii). In [AL, Appendix 1] there is an exposition of the proof of the equivalence of (iv) and (v), and in [KeL, Theorem 4.72] there is also an exposition of the proof of equivalence of (v) with (i) and (ii), in the case of E-invariant μ. For further characterizations, for E-invariant μ, see also [KeL, Section 4.8.2] and [KT, Lemma 3.10]. Also see [Kai] for another equivalent condition, called (IS), for E-quasi-invariant μ, analogous to μ-KE-amenability.

Note that Theorem 8.21, together with Proposition 8.3, implies also Theorem 7.31.

We say that E is **measure amenable** if it is μ-amenable for every probability measure μ. Thus E is measure amenable if and only if it is measure hyperfinite. It turns out that, assuming the Continuum Hypothesis (CH), this is also equivalent to the condition in Remark 8.20 above but in which one requires that the map $x \mapsto \varphi_{[x]_E}(f_x)$ is universally measurable; see [Ke3] and [JKL, 2.8], where other equivalent conditions are also formulated. The role of the CH here comes from the following result used in the proof of this equivalence, due independently to Christensen [Chr] and Mokobodzki (see [DM]).

Theorem 8.22 *Assume the* CH. *Then there is a mean φ on $\mathbb{N}$ such that φ assigns the value 0 to every eventually 0 function, and it is universally measurable in the sense that $\varphi \upharpoonright [-1,1]^{\mathbb{N}} \colon [1,-1]^{\mathbb{N}} \to [-1,1]$ is universally measurable.*

It is now known, see [La], that this result cannot be proved in ZFC. However, Christensen and Mokobodzki have also shown, in ZFC alone, that for each probability measure μ on $[-1,1]^{\mathbb{N}}$, one can find such φ for which $\varphi \upharpoonright [-1,1]^{\mathbb{N}} \colon [-1,1]^{\mathbb{N}} \to [-1,1]$ is μ-measurable.

Remark 8.23 The condition discussed in Remark 8.20 can be easily seen to imply μ-CFW-amenability (see [CFW, page 437]). It can be also shown that, *if one assumes* the CH, then conversely μ-hyperfiniteness implies this condition (see, e.g., [JKL, proof of 2.5(i)]). It seems to be unknown if CH can be eliminated here, and so it seems to be unknown (as pointed out by Tserunyan and Tucker-Drob) whether μ-CFW-amenability and the condition in Remark 8.20 are equivalent (as mentioned in several papers) without using the CH.

8.5 Free Groups and Failure of μ-Amenability

A special case of Proposition 7.30 and Theorem 8.21 shows that if a is a free action of the free group $\mathbb{F}_2 = \langle g_1, g_2 \rangle$, where g_1, g_2 are free generators, on a standard Borel space that admits an invariant probability measure μ, then E_a is not μ-amenable. It is shown in [CG, Théorème 1] and [BG, Section 5] that if a admits a quasi-invariant probability measure μ and each generator g_1, g_2 produces a nonsmooth subequivalence relation of E_a, when restricted to every Borel set of positive μ-measure, then E_a is not μ-amenable. Using this, one can show that if $\mathbb{F}_2 \leq G$, where G is a Lie group and $\mathbb{F}_2$ is not discrete, the translation action of $\mathbb{F}_2$ on G is not μ-hyperfinite, where μ is (a probability measure in the measure class of) the Haar measure; see [CG, Théorème 2]. In [Mo, 4.8] the following generalization is proved:

Theorem 8.24 ([Mo, 4.8]) *Let X be a Polish space and μ be a measure on X such that the open subsets of X of finite measure generate the topology of X. Suppose that a is a free action of $\mathbb{F}_2$ on X, which is continuous with respect to a metrizable nondiscrete topology on $\mathbb{F}_2$. Then E_a is not μ-amenable.*

9

Treeability

9.1 Graphings and Treeings

Let $\mathcal{G}$ be the class of all countable connected graphs and $\mathcal{T} \subseteq \mathcal{G}$ be the class of all countable trees, i.e., connected acyclic graphs. Clearly these are both Borel classes.

Let E be a countable Borel equivalence relation on a standard Borel space X. A **Borel graphing** of E is a $\mathcal{G}$-structure on E, i.e., a Borel graph $\mathbb{G} = \langle X, R \rangle$ with $E = E_{\mathbb{G}}$. A **Borel treeing** of E is a $\mathcal{T}$-structure on E, i.e., a graphing $\mathbb{G}$ that is acyclic. Graphings and treeings of countable Borel equivalence relations play an important role in the study of countable Borel equivalence relations both in the measure-theoretic context, e.g., in the Levitt–Gaboriau theory of costs (see [Le], [Ga1] and [KM1]), and in the Borel-theoretic context. Graphings and treeings in the measure-theoretic context were introduced in [A1], [A3].

Definition 9.1 A countable Borel equivalence relation E is **treeable** if it admits a Borel treeing.

Equivalently, if $\mathcal{T}$ is the class of countable trees, then E is treeable if and only if E is $\mathcal{T}$-structurable. The most obvious examples of treeable countable Borel equivalence relations are generated by free actions of free groups. Let $\mathbb{F}_n$ be the countable free group with $n \leq \infty$ generators S, and let a be a free Borel action of $\mathbb{F}_n$ on a standard Borel space X. Consider the Borel graph $\mathbb{G} = \langle X, R \rangle$, where $xRy \iff$ there exists $s \in S(s \cdot x = y$ or $s \cdot y = x)$. Clearly this is a treeing of E_a. In particular, $F(\mathbb{F}_n, Y)$, for any standard Borel space Y, is treeable, but we will see later that $E_\infty \sim_B E(\mathbb{F}_n, 2)$, $n \geq 2$, is not treeable. Also every hyperfinite Borel equivalence relation is treeable, but there are treeable countable Borel equivalence relations that are not hyperfinite, such as, for example, $F(\mathbb{F}_2, 2)$.

9.2 Equivalent Formulations and Universality

For compressible countable Borel equivalence relations, treeability coincides with generation by free actions of free groups.

Theorem 9.2 ([JKL, 3.16, 3.17]) *Let E be a compressible countable Borel equivalence relation. Then the following are equivalent:*

(i) *E is treeable.*
(ii) *For every free group $\mathbb{F}_n, 2 \leq n \leq \infty$, there is a free Borel action a of $\mathbb{F}_n$ such that $E = E_a$.*
(iii) *For every countable group G such that $\mathbb{F}_2 \leq G$, there is a free Borel action a of G such that $E = E_a$.*

It follows from the theory of cost that there are noncompressible treeable countable Borel equivalence relations that are not generated by free actions of free groups; see, e.g., [KM1, 36.4].

In what follows, we will give a number of equivalent formulations of treeability. We need some definitions first.

Let G be a Polish group, and let a be a Borel action of G on a standard Borel space X. We say that this action has the **cocycle property** if there is a Borel cocycle $\alpha\colon E_a \to G$ such that $\alpha(x, y) \cdot x = y$, see [HK1]. Clearly any free action has this property. We say that a countable Borel equivalence relation is **locally finite treeable** it admits a locally finite treeing. We now have:

Theorem 9.3 ([JKL, 3.3, 3.7, 3.12, 3.17, page 45]) *Let E be a countable Borel equivalence relation. Then the following are equivalent:*

(i) *E is treeable.*
(ii) *E is locally finite treeable.*
(iii) *For any free group $\mathbb{F}_n, 2 \leq n \leq \infty$, there is a free Borel action a of $\mathbb{F}_n$ such that $E \sim_B E_a$.*
(iv) *For every countable group G with $\mathbb{F}_2 \leq G$, there is a free Borel action a of G such that $E \sim_B E_a$.*
(v) *$E \sqsubseteq_B F(\mathbb{F}_2, 2)$.*
(vi) *$E \leq_B F(\mathbb{F}_2, 2)$.*
(vii) *For every countable group G and every Borel action a of G with $E = E_a$, the action a has the cocycle property.*
(viii) *For every countable group G and every Borel action a of G with $E = E_a$, there is a free Borel action b of G with $E \sim_B E_b$.*
(ix) *For every countable group G and every Borel action a of G with $E \subseteq E_a$, there is a free Borel action b of G such that $E \sqsubseteq_B E_b$.*

Further equivalent conditions in the context of groupoids are contained in [L, 8.4.1], which studies functorial Borel complexity for Polish groupoids.

Remark 9.4 The proof of the equivalence of (i) and (ii) in Theorem 9.3 also shows that any countable Borel equivalence relation admits a locally finite graphing, see [JKL, page 50]. For a stronger statement, see Example 7.7(9).

From the theory of cost, it follows that there are noncompressible, treeable countable Borel equivalence relations that do not admit a bounded degree treeing. However, Theorem 9.2(iii) for the group $G = \mathbb{Z}/2\mathbb{Z} \star \mathbb{Z}/2\mathbb{Z} \star \mathbb{Z}/2\mathbb{Z}$ shows that every compressible treeable countable Borel equivalence relation has a Borel treeing in which every vertex has degree 3.

A treeable countable Borel equivalence E is **invariantly universal treeable** if for every treeable countable Borel equivalence relation F, $F \sqsubseteq_B^i E$. As a special case of a general result, see [CK, Corollary 4.4], such a universal equivalence relation exists, and it is of course unique up to Borel isomorphism. It will be denoted by $E_{\infty\mathcal{T}}$. Clearly $E_{\infty\mathcal{T}} \sim_B F(\mathbb{F}_2, 2)$. However, $E_{\infty\mathcal{T}}$ cannot be generated by a free Borel action of a countable group. To see this, notice, following [A2], that $E_0 \oplus F(\mathbb{F}_2, 2)$ cannot be generated by any free Borel action of a countable group (because such a group would have to be amenable by Proposition 7.30).

However, we have the following:

Theorem 9.5 ([H7, Corollary 1.3]) *Let E be a treeable countable Borel equivalence relation on a standard Borel space X, and let μ be an E-ergodic, E-invariant probability measure. Then there are a countable group G, a Borel E-invariant set $Y \subseteq X$ with $\mu(Y) = 1$ and a free Borel action a of G on Y with $E_a = E \upharpoonright Y$.*

This fails for some nontreeable relations by a result of Furman [Fu].

9.3 Closure Properties

The following are the basic closure properties of treeability:

Theorem 9.6 ([JKL, Proposition 3.3]) *Let E, F, E_n be countable Borel equivalence relations. Then we have:*

(i) *If F is treeable and $E \leq_B^w F$, then E is treeable.*
(ii) *If each E_n is treeable, so is $\bigoplus_n E_n$.*

Although the product of a treeable countable Borel equivalence relation with a smooth one is treeable, it is not true that the product of a treeable countable Borel equivalence relation with a hyperfinite one is treeable, and also it is not true that the union of an increasing sequence of treeable countable Borel equivalence relations is treeable, see Theorem 9.12 following.

The following is a basic open problem:

Problem 9.7 Let $E \subseteq F$ be countable Borel equivalence relations such that E is treeable and every F-class contains only finitely many E-classes. Is F treeable?

For several results related to this problem, see [Ts3].

9.4 Essential and Measure Treeability

We say that a Borel equivalence relation E is **reducible-to-treeable** (resp., **essentially treeable**) if it is Borel reducible to a treeable countable Borel equivalence relation (resp., Borel bireducible with a treeable countable Borel equivalence relation). The following is a strengthening of Theorem 3.3.

Theorem 9.8 ([H4], [I2]) *There is a Borel equivalence relation that is reducible-to-treeable but not essentially treeable.*

The proof proceeds by using the results of [I2] to show that the family of equivalence relations E^2_S, in the notation of the paragraph before Theorem 6.8, satisfies all the conditions of [H4, 0.2] (Adrian Ioana pointed out that the proof of [I2, Corollary 4.4, (1)] can be used to show that this family satisfies condition (iii) in [H4, 0.2].)

However, by Theorem 3.6, if E is an idealistic Borel equivalence relation, then the following are equivalent:

(i) E is reducible-to-treeable.
(ii) E is essentially treeable.
(iii) E admits a complete countable Borel section A such that $E \restriction A$ is treeable.

Let E be a countable Borel equivalence relation on a standard Borel space X, and let μ be a probability measure on X. Then we say that E is **μ-treeable** if there is a Borel E-invariant set $A \subseteq X$ such that $\mu(A) = 1$ and $E \restriction A$ is treeable. Finally E is **measure treeable** if it is μ-treeable for every probability measure μ.

Problem 9.9

(i) Is every measure treeable countable Borel equivalence relation treeable?

(ii) Does the analog of Problem 9.7 have a positive answer in the case of μ-treeability, even for F-invariant μ.

9.5 Treeings and μ-Hyperfiniteness

For each tree $\mathbb{T} = \langle T, R \rangle$, two infinite paths $(x_n), (y_n)$ (without backtracking) in T are **equivalent** if there exist m and n such that for all $k (x_{m+k} = y_{n+k})$. The boundary $\partial\mathbb{T}$ of $\mathbb{T}$ is the set of equivalence classes of paths. The following result was proved in [A3] for locally finite treeings and in [JKL] in general.

Theorem 9.10 ([A3], [JKL, Section 3.6]) *Let $\mathbb{G}$ be a Borel treeing of a countable Borel equivalence relation E on a standard Borel space X. For each E-class C, let $\mathbb{T}_C = \mathbb{G} \upharpoonright C$. Let μ be an E-invariant probability measure. Then E is μ-hyperfinite if and only if* $\text{card}(\partial\mathbb{T}_{[x]_E}) \leq 2$, *$\mu$-a.e. (x).*

In fact, it follows from Example 7.7(9) that if $\mathbb{G}$ is a Borel treeing of a countable Borel equivalence relation E on a standard Borel space X and $\text{card}(\partial\mathbb{T}_{[x]_E}) = 2$ for each $x \in X$, then E is hyperfinite and, from Example 7.7(1), it follows that if $\text{card}(\partial\mathbb{T}_{[x]_E}) = 1$ for each $x \in X$, then E is also hyperfinite. Also for any Borel treeing $\mathbb{G}$ of E and *any* probability measure μ on X, $\text{card}(\partial\mathbb{T}_{[x]_E}) \leq 2$, μ-a.e.(x), implies that E is μ-hyperfinite. The converse is not necessarily true if μ is not E-invariant (see Example 7.7(4) or the paragraph following Theorem 10.2).

Gaboriau [Ga1, IV.24], as a corollary of a result of Ghys, generalized Theorem 9.10 to graphings. Tserunyan and Tucker-Drob (see [TT] and [CTT, 1.3]) generalized Theorem 9.10 to E-quasi-invariant measures (for treeings), and Chen, Terlov, and Tserunyan in [CTT, 1.4] generalized the Gaboriau–Ghys result to E-quasi-invariant measures.

It is unknown whether every treeable countable Borel equivalence relation that is also Fréchet amenable is hyperfinite. This is a special case of Problem 8.12.

L. Bowen and Tserunyan–Tucker-Drob have studied the structure of hyperfinite subequivalence relations of μ-treeable countable Borel equivalence relations, see [TT].

9.6 Examples

(1) A construction discussed in [JKL, Section 3.2] shows the following. Let G be a Polish group and $\mathbb{A}$ be a countable structure with universe A for which there is a Borel action of G on A by automorphisms of $\mathbb{A}$ so that the stabilizer of every point in A in this action is compact. Then for every free Borel action a of G on a standard Borel space X, one can find an **$\mathbb{A}$-structurable** countable Borel equivalence relation E (i.e., $\mathcal{K}$-structurable, where $\mathcal{K}$ is the isomorphism class of $\mathbb{A}$) such that $E_a \sim_B E$.

We say that a Polish group G is **strongly Borel treeable** if for every free Borel action a of G on a standard Borel space, E_a is essentially treeable. Using the inducing construction, see Section 2.3, it follows that a closed subgroup of a strongly Borel treeable group is strongly Borel treeable. See also [CGMT, Appendix B] for more closure properties of the class of strongly Borel treeable countable groups. From [SeT, Theorem 1.1] (see also Theorem 10.2) it follows that a countable group G is strongly Borel treeable if and only if $F(G, 2)$ is treeable.

We now have:

Proposition 9.11 ([JKL, Proposition 3.4]) *Let G be a Polish group that has a Borel action on a countable tree with compact stabilizers. Then G is strongly Borel treeable.*

In particular, if a countable group G acts on a countable tree with finite stabilizers, then G is strongly Borel treeable. This includes, for example, groups that contain a free subgroup with finite index, free products of finite cyclic groups, and in particular $\mathrm{PSL}_2(\mathbb{Z}), \mathrm{SL}_2(\mathbb{Z}), \mathrm{GL}_2(\mathbb{Z})$; see [JKL, page 41].

Other examples of strongly Borel treeable groups include $\mathrm{SL}_2(\mathbb{Q}_p)$ and the group of automorphisms $\mathrm{Aut}(\mathbb{T})$ of a locally finite tree $\mathbb{T}$; see [JKL, page 42].

The canonical action of $\mathrm{GL}_2(\mathbb{Z})$ on the two-dimensional torus $\mathbb{R}^2/\mathbb{Z}^2$ is not free, but it still generates a treeable (not hyperfinite) countable Borel equivalence relation, see [JKL, page 42]. It was also shown in [T3, Theorem 5.11] that the action of $\mathrm{GL}_2(\mathbb{Z})$ on $\mathbb{Q}_p \cup \{\infty\}$ by fractional linear transformations generates a treeable but not hyperfinite equivalence relation. Compare this with Example 7.7(3).

(2) Let $\mathcal{K}$ be the class of all rigid locally finite trees, and let $\sigma \in L_{\omega_1\omega}$ be a sentence such that $\mathcal{K} = \mathrm{Mod}(\sigma)$. Then $\cong_\sigma$ is essentially treeable, in fact $\cong_\sigma \sim_B E_{\infty\mathcal{T}}$, see [JKL, page 43] and [HK1, pages 241–242].

(3) In [Ts5, Corollary 1.5], a sufficient condition for treeability of a countable Borel equivalence relation is given, in connection with graphings and Stallings' theorem on ends of groups.

In [CPTT] it is shown that if there are Borel graphings with "tree-like" large scale geometry of a countable Borel equivalence relation E, then E is treeable.

(4) We call a Polish group G **strongly measure treeable** if for every free Borel action a of G on a standard Borel space X and every probability measure μ on X, there is an invariant Borel set $Y \subseteq X$ with $\mu(Y) = 1$ such that $E_a \upharpoonright Y$ is essentially treeable. It is shown in [CGMT] that the isometry group of the hyperbolic plane $\mathbb{H}^2$, $\mathrm{PSL}_2(\mathbb{R})$, $\mathrm{SL}_2(\mathbb{R})$ (and their closed subgroups, in particular surface groups), finitely generated groups with planar Cayley graphs, and elementarily free groups are all strongly measure treeable.

9.7 Conditions Implying Nontreeability

9.7.1 Product Indecomposability

The first obstruction to treeability has to do with indecomposability under products of treeable equivalence relations. The following was proved in [A1] in the locally finite case and in [JKL, 3.27] in general. Another proof can be given using the theory of cost, see, e.g., [KM1, 24.9].

Theorem 9.12 ([A1], [JKL, 3.27]) *Let E_1, E_2 be aperiodic countable Borel equivalence relations, let $E = E_1 \times E_2$, and let μ be an E-invariant probability measure. If E is treeable, then E is μ-hyperfinite.*

In particular it follows that $E_0 \times E_{\infty\mathcal{T}}, E^2_{\infty\mathcal{T}}$ are not treeable and thus in particular

$$E_{\infty\mathcal{T}} <_B E_\infty.$$

Also it follows that $\equiv_T, \equiv_A$ are not treeable.

Recall also here Theorem 6.12, which extends these results and determines the relation under Borel reducibility of $R_n = F(\mathbb{F}_2, 2)^n$ (product of the shifts) and $S_n = F(\mathbb{F}_2^n, 2)$ (shift of the products).

Product indecomposability results for countable Borel equivalence relations generated by free Borel actions of nonamenable hyperbolic groups are contained in [A5, Section 6].

9.7.2 Antitreeable Groups

Let G be a Polish group. We say that G is **antitreeable** if for every free Borel action a of G on a standard Borel space X that admits an invariant probability measure, E_a is not essentially treeable.

Remark 9.13 For a countable group G, being antitreeable means that for every free Borel action a of G, if E_a is not compressible, then it is not treeable. The requirement of noncompressibility is necessary, since every countable group G has a free Borel action a with E_a smooth, therefore treeable.

Theorem 9.14 ([AS, Theorem 1.8]) *Let G be an infinite countable group that has property* (T). *Then G is antitreeable.*

Thus groups such as $\mathrm{SL}_n(\mathbb{Z}), \mathrm{GL}_n(\mathbb{Z}), \mathrm{PSL}_n(\mathbb{Z}), n \geq 3$, are antitreeable. In particular, as opposed to the $n = 2$ case, see Section 9.6(1), the equivalence relation induced by the canonical action of $\mathrm{GL}_n(\mathbb{Z})$ on the torus $\mathbb{R}^n/\mathbb{Z}^n$, for $n \geq 3$, is not treeable.

Actually one has the following strengthening of Theorem 9.14, as noted in [HK1, 10.5]:

Theorem 9.15 ([AS], [HK1, 10.5]) *Let G be an infinite countable group that has property* (T). *Let a be a Borel action of G on a standard Borel space X and let μ be an E_a-invariant, E_a-ergodic probability measure. If F is a treeable, countable Borel equivalence relation, then E_a is μ, F-ergodic.*

Hjorth [H1] used the antitreeability of $\mathrm{SL}_n(\mathbb{Z})$ to show that $\cong_n$, for $n \geq 3$, is not treeable, and this was extended in [Ke7] to the case $n = 2$, see the following analysis. Thomas (unpublished) has shown that $\cong_n^*$, for $n \geq 3$, is not treeable, but it seems to be unknown if this holds in the $n = 2$ case.

Certain products of groups are also antitreeable. Extending results of [Ga1] and [Ke7], the following was shown in [H9].

Theorem 9.16 ([H9, 0.6]) *Let $G = G_1 \times G_2$ be the product of two Polish noncompact but locally compact groups, and assume that G is not amenable. Then G is antitreeable.*

Notice that, by the inducing construction, a lattice in a Polish locally compact antitreeable group is also antitreeable. In particular, $\mathrm{PSL}_2(\mathbb{Z}[1/2])$ is antitreeable, being a lattice in $\mathrm{PSL}_2(\mathbb{R}) \times \mathrm{PSL}_2(\mathbb{Q}_2)$, and this was used in [Ke7] to show that $\cong_2$ is not treeable and in [T3, 5.3] to show that $\cong_2^p$ is not treeable. Hjorth [H1, Section 4] had earlier shown that $\cong_n^p$ is not treeable for $n \geq 3$.

Also, in [C, Theorem 1.1], Calderoni used Theorem 9.14, for $n \geq 5$, and a cocycle superrigidity result for $n = 3, 4$, to show that for $n \geq 3$ the equivalence relation induced by the action of $\mathrm{SO}_n(\mathbb{Q})$ on S^{n-1} is not treeable.

Remark 9.17 A stronger ergodicity type result, related to Theorem 9.15, for actions of product groups is also proved in [Ke7, Theorem 10].

Remark 9.18 In [HK4, Chapters 6,7], various results are proved to the extent that for certain countable groups and free Borel actions with invariant probability measure, the associated countable Borel equivalence relation is not Borel reducible even to a finite product of treeable countable Borel equivalence relations.

9.8 Intermediate Treeable Relations

(1) We have seen that there is a simplest nonsmooth treeable countable Borel equivalence relation, namely E_0, and a most complex one, namely $E_{\infty\mathcal{T}}$. Thus all nonsmooth treeable countable Borel equivalence relations are in the interval $[E_0, E_{\infty\mathcal{T}}]$ in the sense of $\leq_B$. The first problem, already raised in [JKL], was whether this interval is nontrivial, i.e., whether there are **intermediate treeable** countable Borel equivalence relations $E_0 <_B E <_B E_{\infty\mathcal{T}}$. This was answered affirmatively by Hjorth in [H3]. To explain his result we first need a definition.

Let G be a countable group, and let a be a Borel action of G on a standard Borel space X. The action a is **modular** if there is a sequence of countable Borel partitions $(\mathcal{P}_n)$ of X that generates the Borel sets of X and is such that each $\mathcal{P}_n$ is invariant under the action.

We now have the following result:

Theorem 9.19 ([H3, Theorem 3.6]) *Consider the equivalence relation $F(\mathbb{F}_2, 2)$, which is $\sim_B E_{\infty\mathcal{T}}$, and the usual product measure μ on $2^{\mathbb{F}_2}$. Then for any Borel invariant set $A \subseteq 2^{\mathbb{F}_2}$, with $\mu(A) = 1$, and any equivalence relation E_a generated by a modular Borel action a of a countable group G, $F(\mathbb{F}_2, 2) \restriction A \not\leq_B^w E_a$.*

There are free Borel actions a of $\mathbb{F}_2$ with invariant probability measure that are modular (see, e.g., [SlSt]). In fact a countable group G admits a free modular Borel action with invariant probability measure if and only if it is residually finite; see [Ke9, 1.4]. Thus we have:

Corollary 9.20 ([H3]) *There are intermediate treeable countable Borel equivalence relations $E_0 <_B E <_B E_{\infty\mathcal{T}}$.*

If a is a modular action of G on X and b is a modular action of H on Y, the action of $G \times H$ on $X \times Y$ is also modular. Using this and Theorem 9.12, it also follows that there are products of two treeable countable Borel equivalence relations that are incomparable in the sense of $\leq_B$ with $E_{\infty\mathcal{T}}$.

One can find in [H3] and [Ke9] several other characterizations of modularity as well as various examples of modular actions, including the translation action of any countable subgroup $G \leq S_\infty$ on S_∞ and the translation action of $\mathrm{SL}_n(\mathbb{Z})$ on $\mathrm{SL}_n(\mathbb{Z}_p)$.

Let us a call a Borel action a of a countable group **antimodular** if for any modular Borel action b of a countable group, $E_a \not\leq^w_B E_b$. Thus Theorem 9.19 says that the shift action of $\mathbb{F}_2$ restricted to any invariant Borel set of measure 1 is antimodular. This was generalized in [Ke9] by extracting a representation-theoretic condition from the proof of Theorem 9.19 that implies antimodularity.

Let a be a Borel action of a countable group G on a standard Borel space X with invariant probability measure μ. The **Koopman representation** associated to a, in symbols $\boldsymbol{\kappa}^a$, is the unitary representation of G on $L^2(X,\mu)$ defined by $g \cdot f(x) = f(g^{-1} \cdot x)$. Its restriction to the orthogonal of the constant functions $L^2_0(X,\mu) = (\mathbb{C}1)^\perp$ is denoted by $\boldsymbol{\kappa}^a_0$. The **(left) regular representation** of G is the unitary representation λ_G of G on $\ell^2(G)$ defined by $g \cdot f(h) = f(g^{-1}h)$. Also $\boldsymbol{\pi} \prec \boldsymbol{\rho}$ denotes **weak containment** of unitary representations of G, see, e.g., [Ke9, Section 2]. Finally we say that the action a is **tempered** if $\kappa^a_0 \prec \lambda_G$. We now have:

Theorem 9.21 ([Ke9, 3.1]) *Let G be a countable group with $\mathbb{F}_2 \leq G$. If a is a Borel action of G on a standard Borel space with a nonatomic invariant probability measure, and a is tempered, then a is antimodular.*

Several examples of tempered actions are given in [Ke9, Sections 4, 5] These include the action of $\mathrm{SL}_2(\mathbb{Z})$ on the two-dimensional torus, which in view of Section 9.6(1) generates a treeable countable Borel equivalence relation R_2. It turns out also that $R_2 <_B E_{\infty\mathcal{T}}$ (see the paragraph following Theorem 9.25). The action of $\mathrm{SL}_n(\mathbb{Z})$ on the n-dimensional torus is not tempered, if $n \geq 3$, but it is still antimodular (even when restricted to an invariant Borel set of measure 1). This is because there is a copy of $\mathbb{F}_2$ in $\mathrm{SL}_n(\mathbb{Z})$ such that its action on the n-dimensional torus is tempered.

These results were further generalized in [ET]. A unitary representation π of a countable group G on a Hilbert space H is called **amenable (in the sense of Bekka**) if there is a bounded linear functional Φ on the C^*-algebra $B(H)$ of bounded linear operators on H such that $\Phi \geq 0$, $\Phi(I) = 1$ and $\Phi(\pi(g)S\pi(g^{-1})) = \Phi(S)$, for every $g \in G, S \in B(H)$. The relevant point here is that for a nonamenable countable group G, λ_G is not amenable and that if $\pi \prec \rho$ and ρ is not amenable, so is π. Thus if G is not amenable and a is a tempered Borel action of G on a standard Borel space with invariant probability measure, then κ_0^a is not amenable. The following result, which is a corollary of a stronger result proved in [ET], generalizes Theorem 9.21:

Theorem 9.22 ([ET, 1.3]) *Let G be a countable group, and let a be a Borel action of G on a standard Borel space with a nonatomic invariant probability measure. If κ_0^a is nonamenable, then a is antimodular. In particular, Theorem 9.21 is true for any nonamenable group G.*

As a consequence the following is also shown in [ET]:

Theorem 9.23 ([ET, 1.5, 1.6])

(i) *Let the countable group G have property* (T). *Then any probability measure-preserving, weakly mixing action of G is antimodular.*
(ii) *Let the countable group G fail the Haagerup Approximation Property* (HAP). *Then any probability measure-preserving, mixing action of G is antimodular.*

Further antimodularity results are obtained in [I4].

(2) The next question is whether there are uncountably many incomparable, under Borel reducibility, treeable equivalence relations. Inspired by work in [I1], Hjorth provided a positive answer. In fact he showed the following:

Theorem 9.24 ([H12], see also [Mi11, 6.1]) *Let R_2 be the equivalence relation generated by the action of* $\mathrm{SL}_2(\mathbb{Z})$ *on* $\mathbb{R}^2/\mathbb{Z}^2$, *and let μ be the Lebesgue probability measure on $\mathbb{R}^2/\mathbb{Z}^2$. Then there is a family $(E_r)_{r\in\mathbb{R}}$ of countable Borel equivalence relations such that:*

(i) *$E_r \subseteq E_s \subseteq R_2$, if $r \leq s$.*
(ii) *E_r is induced by a free Borel action of $\mathbb{F}_2$.*
(iii) *If $r \neq s$, then there is no μ-measurable reduction from E_r to E_s.*

Recall from Section 9.6(1) that the equivalence relation R_2 of Theorem 9.24 is treeable.

A streamlined version of Hjorth's work that isolated the key ideas was subsequently developed in [Mi11]. This eventually led to the work in [CM1], [CM2] (see also [Mi15]), in which the following concept was introduced.

Let F be a countable Borel equivalence relation on a standard Borel space Y. For any countable Borel equivalence relation E on a standard Borel space X and probability measure μ on X, consider the space of all (partial) weak Borel reductions $f\colon E \restriction A \leq_B^w F$, where $A \subseteq X$ is a Borel set. This carries the pseudometric $d_\mu(f,g) = \mu(D(f,g))$, where if $f\colon A \to Y$ and $g\colon B \to Y$, then $D(f,g) = \{x \in A \cap B \colon f(x) \neq g(x)\} \cup (A \Delta B)$. We also say that E is

μ-nowhere hyperfinite if there is no Borel set $A \subseteq X$ with $\mu(A) > 0$ and $E \upharpoonright A$ hyperfinite.

Then the countable Borel equivalence relation F is called **projectively separable** if for every E, μ as above such that E is μ-nowhere hyperfinite, the pseudometric d_μ is separable.

The following is then shown in [CM1]:

Theorem 9.25 ([CM1, Theorem B, Proposition 2.3.4])

(i) *The equivalence relation R_2 induced by the action of* $\mathrm{SL}_2(\mathbb{Z})$ *on* $\mathbb{R}^2/\mathbb{Z}^2$ *is projectively separable.*
(ii) *If E, F are countable Borel equivalence relations, $E \leq_B^w F$ and F is projectively separable, so is E.*

In particular, since $\mathbb{R}R_2$ is not projectively separable, it follows that $R_2 <_B E_{\infty\mathcal{T}}$.

Note that every measure hyperfinite countable Borel equivalence relation is projectively separable, but the equivalence relation R_2 is not measure hyperfinite. It is now shown in [CM1] that incomparability, and many more of the complexity phenomena for countable Borel equivalence relations that we saw earlier, occur among subequivalence relations of *any* countable Borel equivalence relation that is not measure hyperfinite and projectively separable and treeable, for example, R_2. Here are some of the main results:

Theorem 9.26 ([CM1, Theorem G, Theorem E, Theorem F, Theorem H]) *Let E be a countable Borel equivalence relation on a standard Borel space X that is not measure hyperfinite and is projectively separable and treeable. Then the following hold:*

(i) *There is a family $(E_r)_{r\in\mathbb{R}}$ of pairwise incomparable under measure reducibility $\leq_M$ countable Borel subequivalence relations of E such that $r \leq s \implies E_r \subseteq E_s$.*
(ii) *$\mathbb{R}E \not\leq_M F$, for every Borel subequivalence relation $F \subseteq E$, and in particular $E <_M E_{\infty\mathcal{T}}$.*
(iii) *For some Borel set $A \subseteq X$, if $F = E \upharpoonright A$, then for each $n \geq 1$, $nF <_M (n+1)F$.*
(iv) *If moreover E is aperiodic and the failure of measure hyperfiniteness for E is witnessed by an invariant probability measure, then there is an aperiodic Borel subequivalence relation $F \subseteq E$ such that for every $n \geq 1$, $F \times I_n \sqsubset_M F \times I_{n+1}$.*

In Theorem 9.26, compare (i) with Theorems 9.24 and 2.38; (ii) with Corollary 9.20; (iii) with Theorem 6.20; (iv) with Theorem 2.33.

In [CM2], the authors study the situations under which certain such results hold for subequivalence relations induced by free actions of free groups.

(3) The first *explicit* examples of uncountably many incomparable, under Borel reducibility, treeable countable Borel equivalence relations were constructed in [I2]. As usual, in what follows we view $\mathrm{SL}_2(\mathbb{Z})$ as a dense subgroup of the compact group $H^2_S = \prod_{p \in S} \mathrm{SL}_2(\mathbb{Z}_p)$, for any nonempty set of primes S. Also for any $G \leq \mathrm{SL}_2(\mathbb{Z})$, we let $K_{G,S}$ be the closure of G in H^2_S. Thus $K_{\mathrm{SL}_2(\mathbb{Z}),S} = H^2_S$. We denote by $E_{G,S}$ the equivalence relation induced by the translation action of G on $K_{G,S}$. Thus $E_{\mathrm{SL}_2(\mathbb{Z}),S} = E^2_S$, in the notation of the paragraph before Theorem 6.8.

Theorem 9.27 ([I2, Corollary C]) *Let $S \neq T$ be nonempty sets of primes and $G, H \leq \mathrm{SL}_2(\mathbb{Z})$ be nonamenable. Then $E_{G,S}, E_{H,T}$ are treeable and incomparable in $\leq_B$.*

Recall from Section 9.6(1) that the equivalence relation F_p induced by the action of $\mathrm{GL}_2(\mathbb{Z})$ on $\mathbb{Q}_p \cup \{\infty\}$ by fractional linear transformations is treeable. We now have:

Theorem 9.28 ([I2, Corollary D]) *Let p, q be primes. Then*

$$p = q \iff F_p \leq_B F_q.$$

9.9 Contractible Simplicial Complexes

We will discuss here a higher-dimension generalization of treeability. A **simplicial complex** is a countable set X together with a collection S_k of subsets of X of cardinality $k + 1$, for each $k \in \mathbb{N}$, such that all singletons from X are in S_0 and every subset of an element of S_k of cardinality $m + 1 \leq k + 1$ belongs in S_m. (The elements of S_0 are called vertices, the elements of S_1 are called edges, etc.) We say that a simplicial complex $\mathbb{K}$ is **contractible** if its geometric realization is contractible (see, e.g., [Ch1, Section 2]). We can view each simplicial complex as a structure in some language L and then we let C be the class of all countable contractible simplicial complexes, which is a Borel class. A simplicial complex $\mathbb{K}$ is ***n*-dimensional** if $S_n \neq \emptyset$ but $S_m = \emptyset$ for $m > n$. We denote by C_n the class of all countable contractible n-dimensional simplicial complexes (which is again a Borel class).

For $n = 1$, C_1 coincides with the class $\mathcal{T}$ of all countable trees. Thus C_1-structurability coincides with treeability, and C_n-structurability, $n \geq 2$, can be considered as a higher-dimensional analog of treeability. For example, any equivalence relation induced by a free Borel action of $(\mathbb{F}_2)^n$, for $n \geq 1$, is C_n-structurable. The C_n-structurable countable Borel equivalence relations, in a measure-theoretic context, play an important role in Gaboriau's theory of ℓ^2 Betti numbers, see [Ga2]. From his work it follows that for each $n \geq 1$, if E_{n+1} is induced by a free Borel action of $(\mathbb{F}_2)^{n+1}$ with invariant probability measure, then E_{n+1} is C_{n+1}-structurable but not C_n-structurable; in fact E_{n+1} is not even Borel reducible to a C_n-structurable countable Borel equivalence relation (see also [HK4, Appendix D] and [Ke11, Section 7] here).

Again as a special case of a general result, see [CK, Corollary 4.4], there is an invariantly universal C_n-structurable countable Borel equivalence relation, denoted by $E_{\infty C_n}$ (so that $E_{\infty C_1} = E_{\infty \mathcal{T}}$). Thus $E_{\infty C_n} <_B E_{\infty C_{n+1}}$, for each $n \geq 1$.

We have seen in Section 9.2 that every compressible treeable countable Borel equivalence relation admits a Borel treeing in which every vertex has degree 3. We now have the following higher-dimensional analog:

Theorem 9.29 ([Ch1, Corollary 2]) *Let E be a compressible countable Borel equivalence relation that is C_n-structurable. Then E admits a C_n-structure in which every vertex belongs to exactly $2^{n-1}(n^2 + 3n + 2) - 2$ edges.*

Call a simplicial complex **locally finite** if every vertex belongs to finitely many edges. Then we have:

Theorem 9.30 ([Ch2, Corollary 5]) *Every compressible countable Borel equivalence relation admits a C-structure that is locally finite.*

There are many open problems concerning the class of C_n-structurable Borel equivalence relations, including the following:

Problem 9.31 If the countable Borel equivalence relation F is C_n-structurable, $n \geq 2$, and $E \sqsubseteq_B F$, is E also C_n-structurable?

For this and other open problems, see [Ch1, Section 4].

10

Freeness

10.1 Free Actions and Equivalence Relations

Definition 10.1 A countable Borel equivalence relation E on a standard Borel space X is called **free** if there are a countable group G and a free Borel action a of G such that $E = E_a$.

As in Section 5.3, for any infinite countable group G and any free Borel action a of G, we have that $E_a \sqsubseteq_B^i F(G,\mathbb{R})$. Actually, by [JKL, 5.4], we in fact have that there is a Borel embedding of the action a into $s_{G,\mathbb{N}}$, so $E_a \sqsubseteq_B^i F(G,\mathbb{N})$ and therefore $F(G,\mathbb{R}) \cong_B F(G,\mathbb{N})$. For $G = \mathbb{Z}$ and more generally for any infinite countable group G for which all of its Borel actions generate a hyperfinite relation, we have, as follows from Corollary 7.5, that $F(G,\mathbb{R}) \cong_B F(G,2)$. However, it was shown in [T14, 6.3] that for $G = \mathrm{SL}_3(\mathbb{Z})$ (and other groups) $F(G,2) <_B F(G,3) <_B \cdots <_B F(G,\mathbb{N})$. Apparently it is unknown if this also holds for $G = \mathbb{F}_2$.

Despite this, there is still a close relationship between E_a, for a free action a of G, and $F(G,2)$ in view of the following theorem.

Theorem 10.2 ([SeT, 1.1]) *Let G be an infinite countable group, and let a be a free Borel action of G on a standard Borel space X. Then there is a Borel homomorphism $f : X \to 2^G$ from a to $s_{G,2}$ such that $\overline{f(X)} \subseteq F(2^G)$.*

In particular, $f : E_a \to_B F(G,2)$, and $f \upharpoonright C$ is a bijection of every E_a-class C with the $F(G,2)$-class $f(C)$.

Every aperiodic hyperfinite Borel equivalence relation is clearly free. Moreover, by [DJK, 11.2], for every compressible hyperfinite Borel equivalence relation E and every infinite countable group G, there is a free Borel action a of G with $E = E_a$. In general if E is a compressible countable Borel equivalence relation and $E = E_a$, for a free Borel action a of a countable group G, then for any countable group $H \geq G$, there is a free Borel action b of H such that $E = E_b$, see [DJK, 11.1]. We have seen in Theorem 9.2 that in fact every compressible treeable countable Borel equivalence relation is free, but we have also seen in Section 9.2 that there are treeable countable Borel equivalence relations that are not free.

We say that a countable Borel equivalence relation E is **reducible-to-free** if there is a free countable Borel equivalence relation F such that $E \leq_B F$ and **essentially free** if there is a free countable Borel equivalence relation F such that $E \sim_B F$. It is shown in [JKL, 5.13] that E is reducible-to-free if and only if E is essentially free. Also E is essentially free if and only if $E \times I_{\mathbb{N}}$ is free; see [JKL, 5.11]. Moreover, the class of essentially free countable Borel equivalence relations has the following closure properties.

Proposition 10.3 ([JKL, 5.13]) *Let E, F, E_n be countable Borel equivalence relations. Then we have:*

(i) *If F is essentially free and $E \leq_B^w F$, then E is essentially free.*
(ii) *If E, F, E_n are essentially free, so are $\bigoplus_n E_n$, $E \times F$.*

The following is an open problem.

Problem 10.4 Let $E \subseteq F$ be countable Borel equivalence relations such that each F-class contains only finitely many E-classes. If E is essentially free, is F also essentially free?

From Theorem 9.3 it follows that every treeable countable Borel equivalence relation is essentially free (but there are essentially free countable Borel equivalence relations that are not treeable, see, e.g., Theorem 9.12). The question of whether *every* countable Borel equivalence relation is essentially free was raised in [DJK, Section 11]. It was shown in [T12], using the Popa cocycle superrigidity theory, that this is not the case.

Theorem 10.5 ([T12, 3.9]) *Let E be an essentially free countable Borel equivalence relation. Then there is a countable group G such that $F(G, 2) \not\leq_B E$.*

In the following, by a universal essentially free countable Borel equivalence relation we mean an essentially free countable Borel equivalence relation E such that for every essentially free countable Borel equivalence relation F, we have $F \leq_B E$.

Corollary 10.6 *There is no universal essentially free countable Borel equivalence relation. In particular, E_∞ and $\equiv_T$ are not essentially free.*

It follows, for example from Theorem 6.6, that there are continuum many pairwise incomparable under $\leq_B$ free countable Borel equivalence relations.

It is also shown in [T12, 3.13] that there are continuum many incomparable under $\leq_B$ countable Borel equivalence relations that are not essentially free. In [T14, 5.2] it is shown that if $G = B(m,n)$ is the free m-generator Burnside group of exponent n, then, for sufficiently large odd n, $E(G,2)$ is not essentially free.

Finally, [T12, 6.3] raises the question of whether $\cong_n$, for $n \geq 2$, is essentially free.

If E is a countable Borel equivalence relation on a standard Borel space X and μ is a probability measure on X, we say that E is **μ-free,** resp., **μ-essentially free** if there is an E-invariant Borel set $A \subseteq X$ with $\mu(A) = 1$ such that $E \upharpoonright A$ is free, resp., essentially free.

By Theorem 9.5, every treeable countable Borel equivalence relation E is μ-free for any E-invariant, E-ergodic probability measure μ. In [Fu], examples are given of countable Borel equivalence relations for which this fails. Finally, in [H8] an example is constructed of a countable Borel equivalence relation E on a standard Borel space with E-invariant, E-ergodic probability measure μ such that for every E-invariant Borel set $A \subseteq X$ with $\mu(A) = 1$, $E \upharpoonright A$ is not essentially free.

10.2 Everywhere Faithfulness

We finally consider a weakening of the notion of free action. Let a be an action of a group G on a space X. Then a is called **everywhere faithful** if the action of G on every orbit is faithful, that is, for every $g \neq 1_G$ and every orbit C, there is $x \in C$ such that $g \cdot x \neq x$.

It is shown in [Mi6, Page 1] that for every group G that is not isomorphic to $\mathbb{Z}/2\mathbb{Z} \star \mathbb{Z}/2\mathbb{Z}$, of the form $G = H \star K$, with H, K countable nontrivial groups, every compressible countable Borel equivalence relation is generated by an everywhere faithful action of G. Moreover, as a special case of [Mi6, Theorem 20] (see also [Ke10, 4.1.3]), it is shown that if G_n are nontrivial countable groups that are residually amenable, then every aperiodic countable Borel equivalence relation is generated by an everywhere faithful Borel action of $\star_n G_n$.

11

Universality

11.1 Structural Results

Recall that a countable Borel equivalence relation E is universal if for every countable Borel equivalence relation F, $F \leq_B E$ or equivalently $E \sim_B E_\infty$. We will first discuss some structural properties of such equivalence relations.

The first result shows a stronger property enjoyed by all universal relations.

Theorem 11.1 ([MSS, 3.6]) *Let E be a universal countable Borel equivalence relation. Then for every countable Borel equivalence relation F, $F \sqsubseteq_B E$.*

In [JKL, 6.5, (C)] it was asked whether every universal countable Borel equivalence relation is indivisible, in the sense that in any partition of the space into two disjoint invariant Borel sets, the restriction of the equivalence relation to one of these sets is still universal. In fact a much stronger statement turns out to be true.

Theorem 11.2 ([MSS, 3.1, 3.8])

(i) *Let E be a universal countable Borel equivalence relation on a standard Borel space, and let $f \colon E \to_B \Delta_Y$, for some standard Borel space Y. Then for some $y \in Y$, $E \upharpoonright f^{-1}(y)$ is universal.*

(ii) *Let E be a universal countable Borel equivalence relation on a standard Borel space X, and let $X = \bigsqcup_n X_n$ be a Borel partition of X (where the X_n might not be E-invariant). Then for some n, $E \upharpoonright X_n$ is universal.*

The next result, related to a question in [T12, 3.22], shows that universality is always present in null sets.

Theorem 11.3 ([MSS, 3.10]) *Let E be a universal countable Borel equivalence relation on a standard Borel space X, and let μ be a probability measure* *on X. Then there is an E-invariant Borel set $A \subseteq X$ with $\mu(A) = 0$ such that $E \upharpoonright A$ is still universal.*

An alternative way to prove Theorem 11.2 was developed in [M3, Section 4]. In [M3, Section 4.2], a countably complete ultrafilter U on the σ-algebra of E_∞-invariant Borel sets is constructed, reminiscent of the Martin ultrafilter of invariant under $\equiv_T$ Borel sets (where such a Borel set is in the ultrafilter if and only if it contains a cone of Turing degrees). It is shown that for every set $A \in U$, we have $E_\infty \sim_B E \upharpoonright A$, that is, $E \upharpoonright A$ is also universal. This immediately implies Theorem 11.2(i). Using the definition of U, which involves infinite games, one can also prove Theorem 11.2(ii).

11.2 Manifestations of Universality

In what follows, we call a Borel equivalence relation E **essentially universal countable** if $E \sim_B E_\infty$. We will next discuss universal and essential universal equivalence relations that occur in several areas.

11.2.1 Computability Theory

Slaman and Steel proved that the arithmetical equivalence relation is universal.

Theorem 11.4 (see [MSS, Theorem 2.5]) *The equivalence relation $\equiv_A$ is universal.*

The universality of Turing equivalence is an open problem.

Problem 11.5 Is $\equiv_T$ universal?

Note that Martin's conjecture implies a negative answer to this problem. In fact Martin's conjecture easily implies that we cannot even have $\equiv_T \sim_B 2(\equiv_T)$.

We will next discuss some refinements of Turing equivalence that give universal relations. Consider the Polish space $k^{\mathbb{N}}$, where

$$k \in \{2, 3, \dots n, \dots,\} \cup \{\mathbb{N}\}.$$

Then the group S_∞ acts on $k^{\mathbb{N}}$ by the shift $g \cdot x(n) = x(g^{-1}(n))$ and so does (by restriction) any countable subgroup $G \leq S_\infty$. We denote by $\cong^k_G$ the equivalence relation induced by the shift action of G on $k^{\mathbb{N}}$. Although the following results hold for many other countable groups G, we are primarily interested here in the case where G is the group of all recursive permutations of $\mathbb{N}$, in which case we write $\cong^k_{\text{rec}}$ instead of $\cong^k_G$. In particular, $\cong^2_{\text{rec}}$ is the usual notion of recursive isomorphism of subsets of $\mathbb{N}$.

It was shown in [DK] that $\cong^{\mathbb{N}}_{\text{rec}}$ is universal and in [ACH] that $\cong^{5}_{\text{rec}}$ is universal. Finally this was improved to the following:

Theorem 11.6 ([M3, Theorem 1.6]) *The equivalence relation $\cong^{3}_{\text{rec}}$ is universal.*

Surprisingly the following is still open.

Problem 11.7 Is $\cong^{2}_{\text{rec}}$ universal?

Call a countable Borel equivalence relation E **measure universal** if for every countable Borel equivalence relation F on a standard Borel space X and every probability measure on X, there is an F-invariant Borel set $A \subseteq X$ with $\mu(A) = 1$ such that $F \restriction A \leq_B E$. It is unknown if measure universality implies universality. It is shown in [M3, Theorem 1.7] that $\cong^{2}_{\text{rec}}$ is measure universal. Moreover, it is also shown in that paper that the problem of the universality of $\cong^{2}_{\text{rec}}$ is related to problems in Borel graph combinatorics and leads the author to conjecture that the answer to Problem 11.7 is negative; see the discussion in [M3, pages 5–6].

In [M2], computational complexity refinements of Turing equivalence are shown to be universal, including the following:

Theorem 11.8 ([M2, Theorem 1.1]) *Let $\cong^{P}_{T}$ be polynomial-time Turing equivalence. Then $\cong^{P}_{T}$ is universal.*

Now consider a Borel class $\mathcal{K}$ of countable structures, closed under isomorphism, in a countable relational language L. Then for some sentence σ in $L_{\omega_1\omega}$, we have $\mathcal{K} \cap \text{Mod}_{\mathbb{N}}(L) = \text{Mod}(\sigma)$. Then let $\cong_{\mathcal{K}} = \cong_{\sigma}$ be the isomorphism relation for the structures in $\mathcal{K}$. This is induced by the logic action of S_∞ on $\text{Mod}(\sigma)$. Consider again the restriction of this action to the subgroup of recursive permutations of $\mathbb{N}$, and denote by $\cong^{\text{rec}}_{\mathcal{K}} = \cong^{\text{rec}}_{\sigma}$ the induced equivalence relation, that is, the relation of **recursive isomorphism** for the structures in $\mathcal{K}$. Again, see Remark 3.23, in the case where we consider structures in a language with function symbols, we replace them by their graphs.

In [ACH] and [Ca], various recursive isomorphism relations are shown to be universal, including the following:

Theorem 11.9

(i) [ACH, 3.8] *Let $\mathcal{K}$ be the class of unary functions that are permutations. Then $\cong^{\text{rec}}_{\mathcal{K}}$ is universal. Similarly for the class of equivalence relations.*

(ii) [Ca] *Let $\mathcal{K}$ be one of the following classes: trees, groups, Boolean algebras, fields, linear orders. Then $\cong^{\text{rec}}_{\mathcal{K}}$ is universal.*

In particular, the first part of Theorem 11.9(i) says that the equivalence relation induced by the conjugacy action of the group of recursive permutations on S_∞ is universal.

11.2.2 Isomorphism of Countable Structures

We will now consider the isomorphism relation $\cong_{\mathcal{K}}$ of various classes of structures $\mathcal{K}$.

Theorem 11.10 *The isomorphism relations of the following classes of structures are essentially universal countable:*

(i) ([TV1]) *Finitely generated groups;* ([H11]) *2-generated groups.*
(ii) ([TV2]) *Fields of finite transcendence degree over the rationals.*
(iii) ([JKL, 4.11]) *Locally finite trees.*

Two finitely generated groups G, H are **commensurable** if they have finite index subgroups that are isomorphic. This is a Borel equivalence relation, which is essentially universal countable, see [T10, Theorem 1.1].

11.2.3 Groups

For every countable group G, denote by $E_{\text{conj}}(G)$ the equivalence relation of conjugacy in the (compact metrizable) space of all subgroups of G. In [TV1] it was shown that $E_{\text{conj}}(\mathbb{F}_2)$ is universal. Later it was shown that this holds for all groups containing $\mathbb{F}_2$.

Theorem 11.11 ([ACH, 1.3]) *The equivalence relation $E_{\text{conj}}(G)$ is universal for all countable groups G such that $\mathbb{F}_2 \leq G$.*

In [G2] it was shown that for every countable group G and any nontrivial cyclic group H, $E(G, 2) \leq_B E_{\text{conj}}(G \star H)$, and this gives another proof of Theorem 11.11 for $\mathbb{F}_2$.

Various results about the equivalence relation $E_{\text{conj}}(G)$ are proved in [T14, Section 4]. In particular it is shown in [T14, 4.7] that there are continuum many countable groups G for which the relations $E_{\text{conj}}(G)$ are essentially free and pairwise $\leq_B$-incomparable, thus nonuniversal.

In [TV1, Theorem 8] it is shown that the equivalence relation induced by the action of $\text{Aut}(\mathbb{F}_5)$ on the space of subgroups of $\mathbb{F}_5$ is also universal.

For a left-orderable countable group G, let $\text{LO}(G)$ be the compact metrizable space of all left-orderings of G. The group G acts continuously by conjugation on $LO(G)$, and the corresponding equivalence relation $E_{lo}(G)$ is studied in [CC], where in particular it is shown that $E_{lo}(\mathbb{F}_n)$ is universal, for all $n \geq 2$.

Finally, in a different direction, consider the action of $SL_2(\mathbb{Z})$ on $\mathbb{Z}^2$. This induces a shift action of $SL_2(\mathbb{Z})$ on $2^{\mathbb{Z}^2}$. It is shown in [G1] that this shift action generates a universal relation.

11.2.4 Topological Dynamics

For a countable group G, consider now the isomorphism relation $E^k_{\text{ssh}}(G)$ of subshifts of k^G, $k \geq 2$, which is defined in Example 7.7(6), whose result we recall in the following.

In [Cl2] it was shown that $E^k_{\text{ssh}}(\mathbb{Z}^n)$ is universal. This was extended in [GJS2, 9.4.9] and [Cl3, Section 2] to show the following:

Theorem 11.12 ([GJS2, 9.4.9], [Cl3, Section 2]) *Let G be a countable group that is not locally finite. Then $E^k_{\text{ssh}}(G)$ is universal. The same holds for the restriction of $E^k_{\text{ssh}}(G)$ to free subshifts.*

However, as shown in [GJS2, 9.4.3], $E^k_{\text{ssh}}(G) \sim_B E_0$, if G is an infinite countable group which is locally finite, so we have here a strong dichotomy.

11.2.5 Riemann Surfaces and Complex Domains

Let $\cong_R$ be the isomorphism (conformal equivalence) relation of Riemann surfaces, and let $\cong_D$ be its restriction to complex domains (open connected subsets of $\mathbb{C}$), in an appropriate standard Borel space of parameters for Riemann surfaces and domains, see [HK2, Section 3]. Then we have:

Theorem 11.13 ([HK2, 4.1]) *The equivalence relations $\cong_R$ and $\cong_D$ are essentially universal countable.*

In fact the same holds even if one restricts the isomorphism relation to complex domains of the form $\mathbb{H} \setminus S$, where S is a discrete subset of the upper half plane $\mathbb{H}$. However, we have seen in Section 8.1 that the isomorphism relation on domains of the form $\mathbb{C} \setminus S$, where S is a discrete subset of $\mathbb{C}$, is essentially amenable, so it is not essentially universal countable. It is also shown in [HK2, 5.2] that the conjugacy equivalence relation on the space of discrete subgroups of $PSL_2(\mathbb{R})$ is also essentially universal countable.

11.2.6 Isometric Classification

Recall here Section 3.6. Concerning Theorem 3.28 we actually have the following:

Theorem 11.14 (Hjorth; see [GK, 7.1]) *Let $\mathcal{M}$ be as in Theorem 3.28. Then the relation $\cong^{iso}_{\mathcal{M}}$ is an essentially universal countable Borel equivalence relation.*

11.2.7 Universal Countable Quasi-orders

Let Q be a Borel quasi-order on a standard Borel space X. We say that Q is **countable** if for each $x \in X$ the set $\{y \in X : yQx\}$ is countable. In particular, a countable Borel equivalence relation is a countable quasi-order. With each countable Borel quasi-order Q we associate the countable Borel equivalence relation $xE_Qy \iff xQy \,\&\, yQx$. As with equivalence relations, if Q, Q' are quasi-orders on standard Borel spaces X, X', resp., then we let $f : Q \leq_B Q'$ denote that $f : X \to Y$ is a Borel function such that $xQy \iff f(x)Q'f(y)$, and we say that Q is **Borel reducible** to Q', in symbols $Q \leq_B Q'$, if such an f exists. Note that if $Q \leq_B Q'$, then $E_Q \leq_B E_{Q'}$. Borel reducibility for countable Borel quasi-orders was studied in [W1], where the following analogs of results for countable Borel equivalence relations were proved.

If a is a Borel action of a countable monoid S on a standard Borel space X such that for every $s \in S$, the map $x \mapsto s \cdot x$ is countable-to-1, we let Q_a be the countable Borel quasi-order defined by $x \; Q_a \; y \iff$ there exists $s(s \cdot x = y)$. For a countable monoid S and standard Borel space X, we let $s_{S,X}$ be the **shift action** of S on X^S given by $(s \cdot p)_t = p_{ts}$. Let $Q(S, X) = Q_{s_{S,X}}$. We also let $E(S, X)$ be the associated equivalence relation.

We say that a countable Borel quasi-order Q is **universal** if for every countable Borel quasi-order R we have $R \leq_B Q$. Then E_Q is a universal countable Borel equivalence relation. Finally, let $\mathbb{S}_\infty$ be the free monoid with a countably infinite set of generators.

The following is an analog of Theorem 2.3.

Theorem 11.15 ([W1, 2.1]) *If Q is a countable Borel quasi-order on a standard Borel space X, then there are a countable monoid S and a Borel action a of S on X such that $Q = Q_a$.*

Next we have an analog of E_∞. In the following, let $Q_\infty = Q(\mathbb{S}_\infty, \mathbb{R})$.

Theorem 11.16 ([W1, 2.4]) *The quasi-order Q_∞ is a universal countable Borel quasi-order. In particular $E(\mathbb{S}_\infty, \mathbb{R})$ is a universal countable Borel equivalence relation.*

Finally, we have the following, where we consider the relations as living in the space of finitely generated groups, defined for example in [T8, Section 2]; see also Remark 3.26.

Theorem 11.17 ([W1, 1.6]) *The embeddability quasi-order for finitely generated groups is a universal countable Borel quasi-order. In particular, the bi-embeddability equivalence relation for finitely generated groups is a universal countable Borel equivalence relation.*

In view of Problem 11.5, it is a natural question to ask whether $\leq_T$ is a universal countable Borel quasi-order. However, it was recently shown by P. Lutz and B. Siskind that this is not the case (see [LS, 4.17] and also [HL, 1.7]).

11.2.8 Action Universality

We have seen in Proposition 5.9 that $E(\mathbb{F}_2, 2)$ is universal, and since for any $G \leq H$ we have that $E(G, X) \leq_B E(H, X)$, it follows that for any countable group G that contains a copy of $\mathbb{F}_2$, $E(G, 2)$ is universal. It is unknown if there are any other countable groups for which $E(G, 2)$ is universal. More generally, following [T14], call a countable group G **action universal** if there is a Borel action a of G with E_a universal. Then it is unknown if there are action universal groups that do not contain $\mathbb{F}_2$. Clearly no amenable group can be action universal. It is shown in [T14, 1.6] that there are countable nonamenable groups that are not action universal.

11.2.9 Generators and Invariant Universality

Given a Borel action a of a countable group G on a standard Borel space X and $n \in \{2, 3, \dots, \mathbb{N}\}$, an ***n*-generator** is a Borel partition $X = \bigsqcup_{i<n} A_i$ of X such that $\{g \cdot A_i \colon g \in G, i < n\}$ generates the Borel sets in X. Equivalently such a generator exists if and only if the action a can be Borel embedded into the shift action of G on n^G. It is shown in [JKL, 5.4] that for every Borel action a of a countable group G for which E_a is aperiodic, there is an $\mathbb{N}$-generator. For every equivalence relation E on a set X, let E^{ap} be the **aperiodic part** of E, that is, $E^{\text{ap}} = E \upharpoonright X_E^{\text{ap}}$, where $X_E^{\text{ap}} = \{x \in X \colon [x]_E \text{ is infinite}\}$. If $E = E_a$ as before, then the aperiodic part of a is the action of G on $X_{E_a}^{\text{ap}}$. Then E_a^{ap} is the associated equivalence relation. For the case of the shift action of G on X^G, denote by $E^{\text{ap}}(G, X)$ the aperiodic part of the associated equivalence relation.

Thus we have seen that for any Borel action a of a countable group G, the aperiodic part of a can be embedded in the aperiodic part of the shift action on $\mathbb{N}^G$. As a consequence, $E_a^{\text{ap}} \sqsubseteq_B^i E^{\text{ap}}(G, \mathbb{N})$, and therefore $E^{\text{ap}}(G, \mathbb{N})$ is invariantly universal for all aperiodic countable Borel equivalence relations given by actions of G. In particular, $E^{\text{ap}}(G, \mathbb{N}) \cong_B E^{\text{ap}}(G, \mathbb{R})$.

Because of entropy considerations, even for the group $G = \mathbb{Z}$ it is not the case that every Borel action of $\mathbb{Z}$ with an invariant probability measure admits a finite generator. The following open problem was raised in [We2]: Is it true that every Borel action a of $\mathbb{Z}$ with E_a compressible has a finite generator? In [JKL, 5.7] this question was extended to actions of arbitrary countable groups.

Recall that any Borel action of a countable group on a standard Borel space is Borel isomorphic to a continuous action of the group on a Polish space, so it is enough to consider this problem for continuous actions. In [Ts4], an affirmative answer (with a 32-generator) was obtained for any continuous action of a countable group on a σ-compact Polish space. Moreover, it was shown that for any countable group and *any* continuous action of G on a Polish space with infinite orbits, there is a comeager invariant Borel set on which the action has a 4-generator. Later in [Ho], and by different methods, the original problem of Weiss, that is, the case $G = \mathbb{Z}$, was shown to have a positive answer with a 2-generator. More recently, Hochman and Seward (unpublished) have extended this to arbitrary countable groups and thus have solved Weiss' problem in complete generality for *all* countable groups.

In [FKSV] the question was considered of whether it is true that we have $E^{\mathrm{ap}}(G,\mathbb{R}) \cong_B E^{\mathrm{ap}}(G,2)$. If this happens, then the group G is called **2-adequate**.

Using the result of Hochman and Seward mentioned above, the following was shown:

Theorem 11.18 ([FKSV, 6.0.4]) *Every infinite countable amenable group is 2-adequate.*

This in particular answers in the negative a question of Thomas [T14, Page 391], who asked whether there are infinite countable amenable groups G for which $E(G,\mathbb{R})$ is not Borel reducible to $E(G,2)$.

Moreover, we have:

Theorem 11.19 ([FKSV, 6.0.10, 6.0.11])

(i) *The free product of any countable group with a 2-adequate group that has an infinite amenable factor and thus, in particular, the free groups $\mathbb{F}_n, 1 \leq n \leq \infty$, are 2-adequate.*

(ii) *Let G be n-generated, $1 \leq n \leq \infty$. Then $G \times \mathbb{F}_n$ is 2-adequate. In particular, all products $\mathbb{F}_m \times \mathbb{F}_n$, $1 \leq m, n \leq \infty$, are 2-adequate.*

However, there are groups that are not 2-adequate:

Theorem 11.20 ([FKSV, 6.0.12]) *The group* $\mathrm{SL}_3(\mathbb{Z})$ *is not* 2-*adequate.*

It is not known if there is a characterization of 2-adequate groups.

11.3 Weak Universality and Martin's Conjecture

Definition 11.21 A countable Borel equivalence relation E is called **weakly universal** if for every countable Borel equivalence relation F, $F \leq_B^w E$.

By Theorem 2.37 this is equivalent to stating that there is a universal countable Borel equivalence relation $E' \subseteq E$. In that form an old question of Hjorth, see [ACH, 1.4] or [JKL, 6.5, (A)], asks the following:

Problem 11.22 Is every weakly universal countable Borel equivalence relation universal?

A special case of this question is the following: If E is a countable Borel equivalence relation, $E' \subseteq E$ is a universal countable Borel equivalence relation, and every E-class contains only finitely many E'-classes, is E universal?

We next discuss some examples of weakly universal countable Borel equivalence relations for which it is not known if there are universal:

(i) Since $E(\mathbb{F}_2, 2) \subseteq \cong^2_{\text{rec}}$, $\cong^2_{\text{rec}}$ and therefore also $\equiv_T$ are weakly universal.
(ii) ([TW, 1.4, 1.5]) The isomorphism and biembeddability relations on Kazhdan groups are weakly universal.
(iii) ([W2, 1.4, 1.5]) The isomorphism relation on finitely generated solvable groups of class 3 is weakly universal (thus the same holds for the isomorphism relation on finitely generated amenable groups).

Recall from the paragraph following Problem 11.5 that Martin's conjecture implies that the weakly universal equivalence relation $\equiv_T$ is not universal. In [T11] it was shown that Martin's conjecture has several strong implications concerning weak universality, including the following:

Theorem 11.23 ([T11, 1.2, 4.5]) *Assume Martin's conjecture. Then there are continuum many weakly universal countable Borel equivalence relations that are pairwise incomparable in $\leq_B$.*

Theorem 11.24 ([T11, 1.4]) *Assume Martin's conjecture. Let E be a countable Borel equivalence relation on a standard Borel space X. Then exactly one of the following holds:*

(i) *E is weakly universal.*
(ii) *For every $f\colon \equiv_T \to_B E$, there is a cone of Turing degrees C such that $f(C)$ is contained in a single E-class.*

It is not even known whether (ii) in this result holds unconditionally for $E = E_0$.

Problem 11.25 Is it true that for every $f\colon \equiv_T \to_B E_0$, there is a cone of Turing degrees C such that $f(C)$ is contained in a single E_0-class?

Also, Theorem 11.24 has the following strong ergodicity consequence for weakly universal countable Borel equivalence relations.

Corollary 11.26 ([T11, 3.1]) *Assume Martin's conjecture. Let E be a countable Borel equivalence relation on a standard Borel space X and F be a countable Borel equivalence relation on a standard Borel space Y. Assume that E is weakly universal but F is not. Then for every $f\colon E \to_B F$, there is an E-invariant Borel set $A \subseteq X$ such that $E \upharpoonright A$ is weakly universal and $f(A)$ is contained in a single F-class.*

Thomas [T12, 3.22] raised the question of whether there exists a countable Borel equivalence relation E and an E-invariant, E-ergodic probability measure such that the restriction of E to every E-invariant Borel set of measure 1 is universal. Such E are called **strongly universal**. Martin's conjecture implies a negative answer.

Theorem 11.27 ([T11, 5.4]) *Assume Martin's conjecture. For any countable Borel equivalence relation E and any probability measure μ, there is an E-invariant Borel set A with $\mu(A) = 1$ such that $E \upharpoonright A$ is not weakly universal.*

Finally, it is shown in [T14, 3.4] that, assuming Martin's conjecture, a countable group G has a Borel action a with E_a weakly universal if and only if $E_{\mathrm{conj}}(G)$ is weakly universal.

11.4 Uniform Universality

The concept of uniform universality was introduced in unpublished work of Montalbán, Reimann, and Slaman and extensively developed in [M3] (see also [MSS]).

Suppose X is a standard Borel space and $(\varphi_n)_{n\in\mathbb{N}}$ is a sequence of partial Borel functions $\varphi_n\colon A_n \to X$, with A_n a Borel subset of X that contains the identity function and is closed under composition. Then $(\varphi_n)_{n\in\mathbb{N}}$ generates the countable Borel equivalence relation $R_{(\varphi_n)}$ defined by

$$xR_{(\varphi_n)}y \iff \text{there exist } m, n(\varphi_m(x) = y \mathbin{\&} \varphi_n(y) = x).$$

In this definition we say that $xR_{(\varphi_n)}y$ via the pair (m, n).

For example, if G is a countable group, say $G = \{g_n : n \in \mathbb{N}\}$, a is a Borel action of G on a standard Borel space X, and $\varphi_n(x) = g_n \cdot x$, then $E_a = R_{(\varphi_n)}$.

Also, on $2^{\mathbb{N}}$ if τ_n is the nth Turing functional (which is a Borel partial function on $2^{\mathbb{N}}$), then $\equiv_T = R_{(\tau_n)}$.

If now $E = R_{(\varphi_n)}$, $F = R_{(\psi_n)}$ are given and $f\colon E \to_B F$, then we say that f is a **uniform homomorphism**, with respect to (φ_n), (ψ_n), if there is a function $u\colon \mathbb{N}^2 \to \mathbb{N}^2$ such that if xEy via (m, n), then $f(x)Ff(y)$ via $u(m, n)$.

For example, if a is a Borel action of a countable group G and b is a free Borel action of a countable group H, then a Borel homomorphism of E_a to E_b is uniform (with respect to the Borel functions given by the actions as above) if and only if the cocycle of the action a to H associated to this homomorphism (see Section 6.2) is simply a homomorphism from G to H.

Definition 11.28 A countable Borel equivalence $E = R_{(\varphi_n)}$ is **uniformly universal** (with respect to (φ_n)) if for every countable Borel equivalence relation $F = R_{(\psi_n)}$, there is $f\colon F \leq_B E$ that is uniform (with respect to (φ_n), (ψ_n)).

We now have the following result:

Theorem 11.29 ([M3, Proposition 3.3]) *For every universal countable Borel equivalence relation E, there is a sequence of partial Borel functions (φ_n) such that $E = R_{(\varphi_n)}$ and E is uniformly universal with respect to (φ_n).*

In what follows, we say that an equivalence relation $E = E_a$ generated by a Borel action a of a countable group G is uniformly universal if it is uniformly universal with respect to the Borel functions given by this action.

Theorem 11.30 ([M3, Theorem 1.5, Theorem 3.1]) *For any countable group G, the following are equivalent:*

(i) *G contains a copy of $\mathbb{F}_2$.*
(ii) *There is a Borel action a of G such that E_a is uniformly universal.*
(iii) *For every standard Borel space X with more than one element, $E(G, X)$ is uniformly universal.*
(iv) *$E_{\text{conj}}(G)$ is uniformly universal.*

In fact as pointed out in [M3, page 20], every known proof that a countable Borel equivalence relation E is universal actually shows that E is uniformly universal for an appropriate (φ_n) such that $E = R_{(\varphi_n)}$. A positive answer to the following problem is conjectured in [M3, Conjecture 3.1]:

Problem 11.31 Is it true that for every universal countable Borel equivalence relation E and every (φ_n) such that $E = R_{(\varphi_n)}$, E is uniformly universal with respect to (φ_n)?

However, as pointed out in [M3, page 20], $\equiv_T = R_{(\tau_n)}$ is not uniformly universal. A much more general result on nonuniform universality is proved in [M3, Theorem 3.4], which includes many other equivalence relations in computability theory, including, for example, many–one equivalence on $2^{\mathbb{N}}$. However, the following is an open problem (compare with Problem 11.7):

Problem 11.32 Is $\cong^2_{\text{rec}}$ uniformly universal?

Uniform universality is not preserved on measure-theoretically large sets.

Theorem 11.33 ([M3, Theorem 3.7]) *If $E = R_{(\varphi_n)}$ is a uniformly universal countable Borel equivalence relation on a standard Borel space X and μ is a probability measure on X, then there is a Borel E-invariant set $A \subseteq X$ with $\mu(A) = 1$ such that $E \restriction A$ is not uniformly universal.*

11.5 Inclusion Universality

Recall from Section 7.1 the concept of the Borel inclusion order $\subseteq_B$, where for countable Borel equivalence relations E, F, we put $E \subseteq_B F$ if and only if there is $E' \cong_B E$ with $E' \subseteq F$. This quasi-order is studied in [FKSV]. We say that a countable Borel equivalence relation F is **inclusion universal** if for each countable Borel equivalence relation E on an uncountable standard Borel space, we have that $E \subseteq_B F$.

Proposition 11.34 (Miller) *There exists an inclusion universal countable Borel equivalence relation.*

Proof We will show that $F = E_\infty \times I_{\mathbb{N}}$ works. First notice that F contains a smooth aperiodic countable Borel equivalence relation, so if E is a smooth countable Borel equivalence relation on an uncountable standard Borel space, then there is a Borel isomorphic copy of E contained in F.

So let E be a nonsmooth countable Borel equivalence relation. We can of course assume that $E = E_\infty \restriction Y$, where E_∞ is on the space X and Y is an uncountable Borel E_∞-invariant subset of X. Let $Z = (X \times \mathbb{N}) \setminus (Y \times \{0\})$. Then let $R \subseteq F \restriction Z$ be an aperiodic smooth Borel equivalence relation and put $E' = (F \restriction Y \times \{0\}) \cup R \subseteq F$. We will check that $E \cong_B E'$.

First we have that $E' \cong_B E \oplus R$. Let $A \subseteq Y$ be an E-invariant Borel set such that $E \restriction A \cong_B R$. Then $E' \cong_B E \oplus R \cong_B R \oplus E \restriction (Y \setminus A) \oplus R \cong_B R \oplus E \restriction (Y \setminus A) \cong_B E$. □

11.6 A Picture

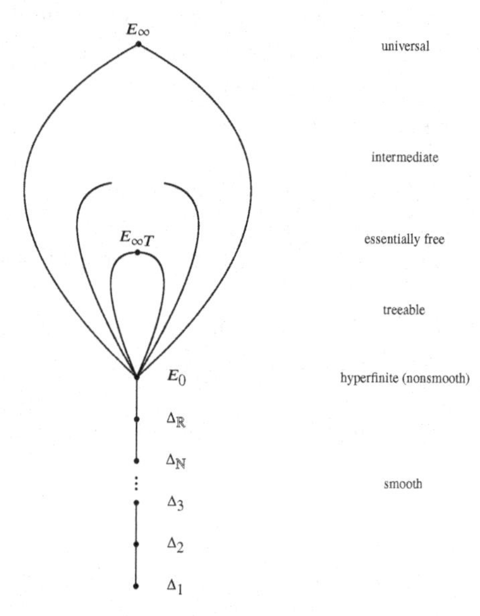

This is a rough picture of the quasi-order of Borel reducibility of countable Borel equivalence relations, in view of the results in Chapters 7 and 9–11. We omit amenability since its extent is not clear at this time.

12

The Poset of Bireducibility Types

Let $\mathcal{E}$ denote the class of countable Borel equivalence relations equipped with the quasi-order $\leq_B$ and the associated equivalence relation $\sim_B$. In this chapter we also allow the empty equivalence relation $\emptyset$ on the empty space as a countable Borel equivalence relation with the convention that $\emptyset \sqsubseteq_B^i E$ for every countable Borel equivalence relation E. For each $E \in \mathcal{E}$ we denote by $\boldsymbol{E} = [E]_{\sim_B}$ the bireducibility class of E, usually called the **bireducibility type** of E. Let $\boldsymbol{\mathcal{E}}$ be the set of bireducibility types. Then $\leq_B$ descends to a partial order $\boldsymbol{E} \leq_B \boldsymbol{F} \iff E \leq_B F$ on $\boldsymbol{\mathcal{E}}$. It has a minimum element $\boldsymbol{\emptyset}$ and a maximum element $\boldsymbol{E}_\infty$. We call $\langle \boldsymbol{\mathcal{E}}, \leq_B \rangle$ the **poset of bireducibility types**. Notice that every bireducibility type $\boldsymbol{e}$ contains a unique Borel isomorphism class consisting of the compressible elements of $\boldsymbol{e}$.

It is clear from Theorem 6.1 that this poset is quite complex, since one can embed in it the poset of Borel subsets of $\mathbb{R}$ under inclusion. Until recently very little was known about the algebraic structure of this poset. Some progress has been now made by applying in this context Tarski's theory of cardinal algebras, see [Ta], which was originally developed as an algebraic approach to the theory of cardinal addition devoid of the use of the Axiom of Choice.

A **cardinal algebra**, see [Ta], is a system $\mathbb{A} = \langle A, +, \sum \rangle$, where $\langle A, + \rangle$ is an abelian semigroup with identity, which will be denoted by 0, and $\sum \colon A^{\mathbb{N}} \to A$ is an infinitary operation satisfying the following axioms, where we put $\sum_{n<\infty} a_n = \sum((a_n)_{n\in\mathbb{N}})$:

(i) $\sum_{n<\infty} a_n = a_0 + \sum_{n<\infty} a_{n+1}$.
(ii) $\sum_{n<\infty}(a_n + b_n) = \sum_{n<\infty} a_n + \sum_{n<\infty} b_n$.
(iii) If $a + b = \sum_{n<\infty} c_n$, then there are $(a_n), (b_n)$ with $a = \sum_{n<\infty} a_n, b = \sum_{n<\infty} b_n, c_n = a_n + b_n$.
(iv) If $(a_n), (b_n)$ are such that $a_n = b_n + a_{n+1}$, then there is c such that for each n, $a_n = c + \sum_{i<\infty} b_{n+i}$.

Let also

$$a \leq b \iff \text{there exists } c(a + c = b).$$

It turns out that this is a partial ordering. All the expected commutativity and associativity laws for $+, \sum$ and monotonicity with respect to $\leq$ hold (see [Ta, Part 1]).

We can define on $\mathcal{E}$ the operations

$$\boldsymbol{E} + \boldsymbol{F} = \text{the bireducibility type of } E \oplus F$$

and

$$\sum_n \boldsymbol{E_n} = \text{the bireducibility type of } \bigoplus_n E_n.$$

In addition, as a special case of a more general result, we now have the following:

Theorem 12.1 ([KMa, 3.3]) *$\langle \mathcal{E}, +, \sum \rangle$ is a cardinal algebra. Moreover, for $\boldsymbol{E}, \boldsymbol{F} \in \mathcal{E}$, $\boldsymbol{E} \leq_B \boldsymbol{F} \iff \boldsymbol{E} \leq \boldsymbol{F}$.*

Also clearly $\boldsymbol{\emptyset}$ is the additive identity of this cardinal algebra. One can now apply the algebraic laws of cardinal algebras established in [Ta] to immediately derive such laws for the poset of bireducibility types and thus for the quasi-order $\leq_B$, including the following:

Theorem 12.2 ([KMa, 1.1])

(i) (Existence of least upper bounds) *Any increasing sequence $F_0 \leq_B F_1 \leq_B \cdots$ of countable Borel equivalence relations has a least upper bound (in the quasi-order $\leq_B$).*

(ii) (Interpolation) *If $\mathcal{S}, \mathcal{T}$ are countable sets of countable Borel equivalence relations and for all $E \in \mathcal{S}$ for all $F \in \mathcal{T}(E \leq_B F)$, then there is a countable Borel equivalence relation G such that for all $E \in \mathcal{S}$ for all $F \in \mathcal{T}(E \leq_B G \leq_B F)$.*

(iii) (Cancellation) *If $n > 0$ and E, F are countable Borel equivalence relations, then*

$$nE \leq_B nF \implies E \leq_B F$$

and therefore

$$nE \sim_B nF \implies E \sim_B F.$$

(iv) (Dichotomy for integer multiples) *For any countable Borel equivalence relation E, exactly one of the following holds:*

1. $E <_B 2E <_B 3E <_B \dots .$

2. $E \sim_B 2E \sim_B 3E \sim_B \dots .$

References to the parts of [Ta] where the relevant laws that are used in Theorem 12.2 are proved can be found in [KMa, 2.2]. Another result proved in [Ta, 3.4] is that in any cardinal algebra, if the infimum (meet) $a \wedge b$ of two elements exists, then the supremum (join) $a \vee b$ exists (and $(a \wedge b) + (a \vee b) = a + b$). It is unknown if the poset of bireducibility types is a lattice, and this can be stated equivalently as follows:

Problem 12.3 Is it true that any two bireducibility types have an infimum?

In fact until very recently it was even unknown if there exist two incomparable under $\leq$ bireducibility types that have an infimum. A positive answer is given in Theorem 13.9.

Finally, concerning the cancellation law for sums, Theorem 12.2(iii), it is natural to ask if there is a similar cancellation law for products. Using methods of ergodic theory it can be shown that this is not the case.

Theorem 12.4 ([KMa, 4.1]) *There are two countable Borel equivalence relations E, F with $E <_B F$ such that $E^2 \sim_B F^2$.*

Remark 12.5 Cardinal algebras also occur in another context in the theory of countable Borel equivalence relations. Let E be a compressible countable Borel equivalence relation on a standard Borel space X. Consider the space of all Borel subsets of X modulo $\sim_E$ (see the paragraph following Definition 2.21). It was shown in [Ch2] that, with some natural operations, this becomes a cardinal algebra, which exhibits interesting properties and is studied in detail in [Ch2].

13

Structurability

13.1 Universal Structurability

We will consider in this section $\mathcal{K}$-structurable countable Borel equivalence relations for Borel classes $\mathcal{K}$ of countable structures in some language L (see Section 8.3). We denote by $\mathcal{E}_{\mathcal{K}}$ the class of countable Borel equivalence relations that are $\mathcal{K}$-structurable. Examples of such classes, for various $\mathcal{K}$, include the following: aperiodic, smooth, compressible, hyperfinite, treeable, α-amenable, the equivalence relations induced by a free Borel action of a fixed countable group G, and all countable Borel equivalence relations.

The next result shows that $\mathcal{E}_{\mathcal{K}}$ contains an invariantly universal element. It was proved in [KST, 7.1] for classes of graphs and generalized by Miller. In what follows, for any class C of countable Borel equivalence relations, a relation $E \in C$ is called **invariantly universal** for C if for $F \in C$, $F \sqsubseteq_B^i E$.

Theorem 13.1 (see [M3, Theorem 4.13], [CK, 1.1]) *For each Borel class $\mathcal{K}$ of countable structures, there is a (unique up to Borel isomorphism) invariantly universal equivalence relation in $\mathcal{E}_{\mathcal{K}}$.*

This invariantly universal relation will be denoted by $E_{\infty\mathcal{K}}$. For example, for the class $\mathcal{E}_{\mathcal{K}}$ of hyperfinite equivalence relations (resp., treeable, induced by a free Borel action of a countable group G, all countable Borel equivalence relations), $E_{\infty\mathcal{K}}$ is Borel isomorphic to $E_{\infty h}$ (resp., $E_{\infty\mathcal{T}}$, $F(G,\mathbb{R})$, E_∞).

13.2 About Universally Structurable Relations

Results that are analogous to those for E_∞ in Section 11.1 have been proved in [M3, Sections 4.3, 4.4] for certain $E_{\infty\sigma}$.

The following is an analog of Theorem 11.3.

Theorem 13.2 ([M3, Theorem 4.4]) *Let σ be a theory in $L_{\omega_1\omega}$. Let $E_{\infty\sigma}$ be on the space $X_{\infty\sigma}$, and let μ be a probability measure on $X_{\infty\sigma}$.Then there is an $E_{\infty\sigma}$-invariant Borel set $A \subseteq X_{\infty\sigma}$ with $\mu(A) = 0$ such that $E_{\infty\sigma} \cong_B E_{\infty\sigma} \upharpoonright A$.*

Recall that a family $(E_i)_{i\in I}$ of equivalence relations on a set X is **independent** if for any sequence $x_0, x_1, \ldots, x_n = x_0$, with $n > 1$, if

$$x_0 E_{i_0} x_1 E_{i_1} x_2 \ldots x_{n-1} E_{i_{n-1}} x_0,$$

where $i_k \neq i_{k+1}$ if $k < n-2$, and $i_{n-1} \neq i_0$, there is $j < n$, with $x_j = x_{j+1}$. In this case we call $\bigvee_i E_i$ an **independent join**.

Note, for example, that the classes $C_n, n \geq 1$, defined in Section 9.9 (which are of the form $\mathcal{E}_{\sigma_n}$ for an appropriate σ_n) are closed under countable independent joins. For σ such that $\mathcal{E}_\sigma$ is closed under independent joins of two relations, it was shown in [M3, Theorem 4.5], generalizing the result mentioned in Section 11.1, that there is a countably complete ultrafilter U on the $E_{\infty\sigma}$-invariant Borel sets such that if $A \in U$, $E_{\infty\sigma} \sim_B E_{\infty\sigma} \upharpoonright A$. From this, and the definition of U, we have the analog of Theorem 11.2:

Theorem 13.3 ([M3, Theorem 4.5, Theorem 4.6]) *Let σ be a theory in $L_{\omega_1\omega}$.*

(i) *Assume that $\mathcal{E}_\sigma$ is closed under independent joins of two relations. Then if $f\colon E \to_B \Delta_Y$, for some standard Borel space Y, there is $y \in Y$ such that $E_{\infty\sigma} \upharpoonright f^{-1}(\{y\}) \sim_B E_{\infty\sigma}$.*
(ii) *If, moreover, $\mathcal{E}_\sigma$ is closed under countable independent joins, $E_{\infty\sigma}$ is on the space $X_{\infty\sigma}$, and $X_{\infty\sigma} = \bigsqcup_n X_n$ is a Borel partition (where the X_n might not be $E_{\infty\sigma}$-invariant), then for some n, $E_{\infty\sigma} \sqsubseteq_B E_{\infty\sigma} \upharpoonright X_n$. In particular, if $E_{\infty\sigma} \leq_B F$, for some countable Borel equivalence relation F, then $E_{\infty\sigma} \sqsubseteq_B F$.*

13.3 Elementary Classes of Relations

By the result of Lopez–Escobar in [LE], a class $\mathcal{K}$ of countable structures is Borel if and only if there is a countable $L_{\omega_1\omega}$ theory, that is, an $L_{\omega_1\omega}$ sentence σ, such that $\mathcal{K}$ is exactly the class of countable models of σ. We thus often write

$$\mathcal{E}_\sigma = \mathcal{E}_\mathcal{K}, E_{\infty\sigma} = E_{\infty\mathcal{K}},$$

and for a countable Borel equivalence relation E, we put

$$E \models \sigma \iff E \in \mathcal{E}_\mathcal{K}.$$

We then say that a class C of countable Borel equivalence relations is **elementary** if $C = \mathcal{E}_\sigma$ for some σ. Thus all the examples we mentioned in Section 13.1 are elementary. The elementary classes can be characterized as follows.

A Borel homomorphism $f : E \to_B F$ between countable Borel equivalence relations E, F on standard Borel spaces X, Y, resp., is called **class bijective** if for each $x \in X$, $f \upharpoonright [x]_E$ is a bijection of $[x]_E$ onto $[f(x)]_F$. We write in this case $f : E \to_B^{cb} F$, and if such f exists we put $E \to_B^{cb} F$. We now have:

Theorem 13.4 ([CK, 1.2]) *Let C be a class of countable Borel equivalence relations. Then C is an elementary class if and only if it is closed downward under $\to_B^{cb}$ and contains an invariantly universal element.*

The following classes are not elementary (see [CK, Section 3.1]): nonsmooth, noncompressible, free, essentially free.

Every countable Borel equivalence relation is contained in a smallest, under inclusion, elementary class.

Theorem 13.5 ([CK, 1.3]) *Let E be a countable Borel equivalence relation. Then $\mathcal{E}_E = \{F \in \mathcal{E} : F \to_B^{cb} E\}$ is the smallest elementary class containing E.*

We next consider elementary classes closed downward under Borel reductions, such as, for example, hyperfinite or treeable. These are called **elementary reducibility classes**.

Theorem 13.6 ([CK, 1.4]) *Let C be an elementary class of countable Borel equivalence relations. Then $C^r = \{F \in \mathcal{E} : \text{there exists } E \in C(F \le_B E)\}$ is the smallest elementary reducibility class containing C.*

Elementary reducibility classes can be characterized as follows. A Borel homomorphism $f : E \to_B F$ between countable Borel equivalence relations E, F on standard Borel spaces X, Y, resp., is called **smooth** if for each $y \in Y$, $E \upharpoonright f^{-1}(\{y\})$ is smooth. This notion was considered in [CCM]. We write in this case $f : E \to_B^{sm} F$, and if such f exists we put $E \to_B^{sm} F$. We now have:

Theorem 13.7 ([CK, 1.5]) *Let C be a class of countable Borel equivalence relations. Then C is an elementary reducibility class if and only if it is closed downward under $\to_B^{sm}$ and contains an invariantly universal element.*

There is an interesting connection between these concepts and the amenability of groups. For each infinite countable group G, let $\mathcal{E}_G^*$ be the elementary class of all countable Borel equivalence relations whose aperiodic part is generated by a free Borel action of G.

Theorem 13.8 ([CK, 1.6]) *Let G be an infinite countable group. Then the following are equivalent:*

(i) *G is amenable.*
(ii) *$\mathcal{E}_G^*$ is an elementary reducibility class.*

We call any countable Borel equivalence relation that is Borel isomorphic to one of the form $E_{\infty\sigma}$, for an $L_{\omega_1\omega}$ theory σ, **universally structurable**. Let $\mathcal{E}_\infty$ be the class of these equivalence relations, and let $\boldsymbol{\mathcal{E}}_\infty = \{\boldsymbol{E} : E \in \mathcal{E}_\infty\}$. Then $\langle \boldsymbol{\mathcal{E}}_\infty, \leq \rangle$ is a subposet of $\langle \boldsymbol{\mathcal{E}}, \leq \rangle$. It is quite rich since it can be shown that the poset of Borel subsets of $\mathbb{R}$ under inclusion can be embedded into it, see [CK, 1.9]. However this subposet has desirable algebraic properties.

Theorem 13.9 ([CK, 1.8]) *The poset $\langle \boldsymbol{\mathcal{E}}_\infty, \leq \rangle$ is a countably complete, distributive lattice. Moreover, the countable meets and joins in this lattice are also meets and joins in the poset $\langle \boldsymbol{\mathcal{E}}, \leq \rangle$.*

Remark 13.10 Notice that if $E \in \mathcal{E}_\infty$, then $\mathbb{R}E \cong_B E$. It follows that $\boldsymbol{\mathcal{E}}_\infty$ is a proper subset of $\boldsymbol{\mathcal{E}}$, even when restricted to nonsmooth countable Borel equivalence relations, see Theorem 6.20. It is unknown if there is $E \notin \mathcal{E}_\infty$ with $\mathbb{R}E \cong_B E$.

It is an interesting problem to understand the connection between the model-theoretic properties of a theory σ and the Borel-theoretic properties of the class $\mathcal{E}_\sigma$. The following result, answering a question of Marks [M3, end of Section 4.3], is a step in that direction.

Theorem 13.11 ([CK, 1.10]) *Let σ be a theory in $L_{\omega_1\omega}$. Then the following are equivalent:*

(i) *Every equivalence relation in $\mathcal{E}_\sigma$ is smooth.*
(ii) *There is a formula $\phi(x)$ in $L_{\omega_1\omega}$ that defines a finite nonempty set in every countable model of σ.*

The next step would be to characterize the σ for which every equivalence relation in $\mathcal{E}_\sigma$ is hyperfinite. However, this is an open problem.

Problem 13.12 Is there a characterization of the $L_{\omega_1\omega}$ theories σ for which every equivalence relation in $\mathcal{E}_\sigma$ is hyperfinite?

It is also of interest to find out which theories σ have the property that every aperiodic countable Borel equivalence relation is in $\mathcal{E}_\sigma$.

Problem 13.13 Is there a characterization of the $L_{\omega_1\omega}$ theories σ for which every aperiodic countable Borel equivalence relation is in $\mathcal{E}_\sigma$?

In the case where σ is the Scott sentence of a countable structure $\mathbb{A}$, the following result was proved by Marks as a consequence of the work in [AFP].

Recall that a countable Borel equivalence relation is $\mathbb{A}$-structurable if and only if it is in $\mathcal{E}_\sigma$ for the Scott sentence σ of $\mathbb{A}$.

Theorem 13.14 (Marks, see [CK, 1.11]) *Let $\mathbb{A}$ be a countable structure with trivial definable closure. Then every aperiodic countable Borel equivalence relation is $\mathbb{A}$-structurable.*

Such structures $\mathbb{A}$ include, for example, many Fraïssé structures, such as the rational order and the random graph, see [CK, 8.17].

14

Topological Realizations

The main concern of [FKSV] is the subject of well-behaved, in some sense, realizations of countable Borel equivalence relations. Generally speaking a realization of a countable Borel equivalence relation E is a countable Borel equivalence relation $F \cong_B E$ with desirable properties.

To start with, a **topological realization** of a countable Borel equivalence relation E on a standard Borel space is an equivalence relation F on a *Polish space* Y such that $E \cong_B F$, in which case we say that F is a topological realization of E in the space Y. It is clear that every E admits a topological realization in some Polish space, but we will look at topological realizations that have additional properties.

Also, by the Feldman–Moore theorem (Theorem 2.3), it is clear that every countable Borel equivalence relation E admits a topological realization, in some Polish space Y, which is induced by a *continuous action* of some countable (discrete) group G on Y. We will look again at such **continuous action realizations** for which the space and the action have additional properties.

To avoid uninteresting situations, unless it is otherwise explicitly stated or clear from the context, in what follows, all the standard Borel or Polish spaces will be uncountable and all countable Borel equivalence relations will be aperiodic. We will denote by $\mathcal{AE}$ the class of all aperiodic countable Borel equivalence relations on uncountable standard Borel spaces.

Concerning topological realizations, we first have the following:

Theorem 14.1 ([FKSV, Theorem 1.1.1]) *For every equivalence relation $E \in \mathcal{AE}$ and every perfect Polish space Y, there is a topological realization of E in Y in which every equivalence class is dense.*

This has in particular as a consequence a stronger version of the marker lemma, see Theorem 2.15. Let E be a countable Borel equivalence relation on a standard Borel space X. A **Lusin marker scheme** for E is a family $\{A_s\}_{s\in\mathbb{N}^{<\mathbb{N}}}$ of Borel sets such that

(i) $A_\emptyset = X$.
(ii) $\{A_{sn}\}_n$ are pairwise disjoint and $\bigsqcup_n A_{sn} \subseteq A_s$.
(iii) Each A_s is a complete section for E.

We have two types of Lusin marker schemes:

(1) The Lusin marker scheme $\{A_s\}_{s\in\mathbb{N}^{<\mathbb{N}}}$ for E is of **type I** if in (ii) above we actually have that $\bigsqcup_n A_{sn} = A_s$ and moreover the following holds:

(iv) For each $x \in \mathbb{N}^\mathbb{N}$, $\bigcap_n A_{x|n}$ is a singleton.

(Then in this case, for each $x \in \mathcal{N}$, $A_n^x = A_{x|n} \setminus \bigcap_n A_{x|n}$ is a vanishing sequence of markers (i.e., $\bigcap_n A_n^x = \emptyset$).)

(2) The Lusin marker scheme $\{A_s\}_{s\in\mathbb{N}^{<\mathbb{N}}}$ for E is of **type II** if it satisfies the following:

(iv′) If for each n, $B_n = \bigsqcup\{A_s : s \in \mathbb{N}^n\}$, then $\{B_n\}$ is a vanishing sequence of markers.

We now have:

Theorem 14.2 ([FKSV, Theorem 1.1.2]) *Every $E \in \mathcal{AE}$ admits a Lusin marker scheme of type* I *and a Lusin marker scheme of type* II.

We next look at continuous action realizations. An important case of such a realization of $E \in \mathcal{AE}$ would be a continuous action realization F on a compact Polish space, called a **compact action realization**. If in addition the action is minimal, we call it a **minimal, compact action realization**. Excluding the case of smooth relations, for which compact action realizations are impossible, we have the following result. In the following, for each countable group G and topological space X, a **subshift** of X^G is the restriction of the shift action of G on X^G to a nonempty shift-invariant closed set.

Theorem 14.3 ([FKSV, Theorem 1.1.3]) *Every nonsmooth hyperfinite equivalence relation in $\mathcal{AE}$ has a minimal, compact action realization. In fact this realization can be taken to be a subshift of $2^{\mathbb{F}_2}$ if the equivalence relation is compressible and a subshift of $2^\mathbb{Z}$ otherwise.*

We next discuss other cases of countable Borel equivalence relations that admit such realizations. Recall that every equivalence relation induced by a free Borel action of a countable group G is Borel isomorphic to the restriction of $F(G, 2^{\mathbb{N}})$ to an invariant Borel set, and similarly every equivalence relation induced by an aperiodic Borel action of G is Borel isomorphic to the restriction of $E^{\mathrm{ap}}(G, 2^{\mathbb{N}})$ to an invariant Borel set. In contrast to Theorem 14.3, the next results show that some very complex countable Borel equivalence relations have compact action realizations.

Theorem 14.4 ([FKSV, Theorem 1.1.4])

(i) *For every infinite countable group G, $F(G, 2^{\mathbb{N}})$ admits a compact action realization. If G is also finitely generated, then $E^{\mathrm{ap}}(G, 2^{\mathbb{N}})$ admits a compact action realization. In fact in both cases such a realization can be taken to be a subshift of $(2^{\mathbb{N}})^G$.*
(ii) *Every compressible, universal countable Borel equivalence relation admits a compact action realization. In fact such a realization can be taken to be a subshift of $2^{\mathbb{F}_4}$.*

In particular, it follows that arithmetical equivalence $\equiv_A$ on $2^{\mathbb{N}}$ has a compact action realization, but it is unknown if Turing equivalence $\equiv_T$ has such a realization.

Problem 14.5 Does Turing equivalence $\equiv_T$ have a compact action realization?

More generally, we do not know whether *every* nonsmooth countable Borel equivalence relation has a compact action realization.

Problem 14.6 Does every nonsmooth aperiodic countable Borel equivalence relation have a compact action realization?

We also do not know if *every* nonsmooth aperiodic countable Borel equivalence relation even admits some other kinds of realizations, for example transitive (i.e., having at least one dense orbit) continuous action realizations on arbitrary or special types of Polish spaces.

In relation to Theorem 14.4, call a countable group G **minimal subshift universal** if there is a minimal subshift of 2^G on which the restriction of the shift equivalence relation is universal. Then we have the following result:

Theorem 14.7 ([FKSV, Theorem 1.2.1]) *Let G and H be infinite countable groups, where H admits a Borel action on a standard Borel space whose induced equivalence relation is universal (e.g., any group containing $\mathbb{F}_2$). Then we have that the wreath product $G \wr H$ is minimal subshift universal. In particular, $\mathbb{F}_3$ is minimal subshift universal.*

The preceding considerations have some connections to the concept of tests of amenability. It is well known that a countable group G is amenable if and only if every continuous action of G on a compact space admits an invariant Borel probability measure. Call a class $\mathcal{F}$ of such actions a **test for amenability** for G if G is amenable provided that every action in $\mathcal{F}$ admits an invariant Borel probability measure. In [GdlH] it is shown that the class of actions on $2^{\mathbb{N}}$ is a test for amenability for any group. Equivalently this says that the class of all subshifts of $(2^{\mathbb{N}})^G$ is a test of amenability for G. It turns out that the strongest result along these lines is actually true, namely that the class of all subshifts of 2^G is a test of amenability for G. This gives a new characterization of amenability.

Theorem 14.8 ([FKSV, Theorem 1.2.2]) *Let G be a countable group. Then G is amenable if and only if every subshift of* 2^G *admits an invariant Borel probability measure.*

Finally, we mention that, in response to a question raised by Conley, it is shown in [FKSV, Section 1.3] that every every $E \in \mathcal{AE}$ admits a K_σ realization on $2^{\mathbb{N}}$ with at least one dense class.

15

A Universal Space for Actions and Equivalence Relations

In [FKSV] the authors also consider a universal space for actions and equivalence relations and study the descriptive or topological properties of various subclasses.

For a countable group G and Polish space X, define the standard Borel space of subshifts of X^G by:

$$\mathrm{Sh}(G, X) = \{F \in F(X^G) : F \text{ is } G\text{-invariant}\}.$$

Here $F(Y)$ is the standard Borel space of nonempty closed subsets of a Polish space Y. If X is compact, we view this as a compact Polish space with the Vietoris topology.

Every compact Polish space is (up to homeomorphism) a closed subspace of the Hilbert cube $\mathbb{I}^{\mathbb{N}}$, and thus every continuous action of a countable group G on a compact Polish space is (topologically) isomorphic to a subshift of $(\mathbb{I}^{\mathbb{N}})^G$. We can thus consider the compact Polish space $\mathrm{Sh}(G, \mathbb{I}^{\mathbb{N}})$ as the universal space of such actions.

Similarly consider the product space $\mathbb{R}^{\mathbb{N}}$. Every Polish space is (up to homeomorphism) a closed subspace of $\mathbb{R}^{\mathbb{N}}$, and thus every continuous action of G on a Polish space is (topologically) isomorphic to a subshift of $(\mathbb{R}^{\mathbb{N}})^G$. We can thus consider the standard Borel space $\mathrm{Sh}(G, \mathbb{R}^{\mathbb{N}})$ as the universal space of such actions.

In particular, taking $G = \mathbb{F}_\infty$, we see that every countable Borel equivalence relation is Borel isomorphic to the equivalence relation E_F induced on some subshift F of $(\mathbb{R}^{\mathbb{N}})^{\mathbb{F}_\infty}$, and so we can view $\mathrm{Sh}(\mathbb{F}_\infty, \mathbb{R}^{\mathbb{N}})$ also as the universal space of countable Borel equivalence relations and study the complexity of various classes of countable Borel equivalence relations (e.g., smooth, aperiodic, hyperfinite) as subsets of this universal space. Similarly we can view $\mathrm{Sh}(\mathbb{F}_\infty, \mathbb{I}^{\mathbb{N}})$ as the universal space of countable Borel equivalence relations that admit a compact action realization. In this case one can also consider both complexity and genericity questions of the various classes. We now state some of the main results proved in [FKSV]. In what follows, G is a countable infinite group. A countable group G is called **exact** if it admits a topologically amenable action on a compact Polish space (see [BO, Chapter 5]). For example, all free groups and all amenable groups are exact.

Theorem 15.1 ([FKSV, Section 4])

(i) *In the space of subshifts of* $(\mathbb{I}^{\mathbb{N}})^G$*: finiteness and smoothness are meager properties; freeness (of the action), aperiodicity, and (for nonamenable* G*) compressibility are comeager. Also (for exact groups* G*) amenability and measure hyperfiniteness are comeager.*
(ii) *In the space of subshifts of* $(\mathbb{I}^{\mathbb{N}})^G$*: (for residually finite* G*) finiteness and smoothness are* $\mathbf{\Pi}^1_1$*-complete; freeness (of the action) and aperiodicity are* G_δ*, and compressibility is open. Also (for residually finite and nonamenable* G*) measure hyperfiniteness is* $\mathbf{\Pi}^1_1$*-complete and hyperfiniteness and amenability are* $\mathbf{\Pi}^1_1$*-hard.*
(iii) *In the space of subshifts of* $(\mathbb{R}^{\mathbb{N}})^G$*: (for residually finite* G*) finiteness and smoothness are* $\mathbf{\Pi}^1_1$*-complete; freeness (of the action), aperiodicity, and compressibility are* $\mathbf{\Pi}^1_1$*-complete; (for nonamenable* G*) hyperfiniteness and amenability are* $\mathbf{\Pi}^1_1$*-hard and measure hyperfiniteness is* $\mathbf{\Pi}^1_1$*-complete.*

The following are open problems:

Problem 15.2 Is hyperfiniteness comeager in $\mathrm{Sh}(G, \mathbb{I}^{\mathbb{N}})$, for exact G?

It was shown recently in [IS] that the answer is positive for the free groups and in fact all groups with finite asymptotic dimension.

Theorem 15.3 ([IS]) *Hyperfiniteness is comeager in* $\mathrm{Sh}(G, \mathbb{I}^{\mathbb{N}})$ *for all groups with finite asymptotic dimension.*

It is unknown whether this holds for every exact group or even every amenable group. Since for exact groups measure hyperfiniteness is comeager, this has some relevance to Problem 7.29.

Problem 15.4 What is is the exact descriptive complexity of hyperfiniteness in the space $\mathrm{Sh}(G, \mathbb{I}^{\mathbb{N}})$?

In connection with Problem 15.2, it is shown in [FKSV, Proposition 4.3.13] that hyperfiniteness is dense in $\mathrm{Sh}(G, \mathbb{I}^{\mathbb{N}})$ for several groups, including all subgroups of hyperbolic groups.

16

Open Problems

We collect here many of the open problems discussed earlier.

16.1 Bireducibility vs. Isomorphism of Quotient Spaces

Problem (2.34) Is it true that $E \sim_B F \iff E \cong^q_B F$?

16.2 Essential Countability and Countable Sections

Problem (3.16) Let G be a Polish group with the property that all the equivalence relations induced by Borel actions of G on standard Borel spaces are Borel and essentially countable. Is the group locally compact?

Problem (3.18) Is it true that the following are equivalent for a Borel equivalence relation E on a standard Borel space X?

(i) E admits a complete countable Borel section.
(ii) There is a Borel assignment $x \mapsto \mu_x$ of probability Borel measures to points $x \in X$ such that $\mu_x([x]_E) = 1$ and $xEy \implies \mu_x \sim \mu_y$.

16.3 Isomorphism of Models of First-Order Theories

Problem (3.27) Let σ be a *first-order* theory, that is, the conjunction of countably many first-order sentences. Is it possible for $\cong_\sigma$ to be Borel, nonsmooth, and essentially countable?

16.4 Hyperfiniteness

Problem (7.24) Let $E_n, n \in \mathbb{N}$, be hyperfinite Borel equivalence relations such that $E_n \subseteq E_{n+1}$ for every n. Is $\bigcup_n E_n$ hyperfinite?

Problem (7.25) What is the extent of the class of Borel-bounded countable Borel equivalence relations? Are they all hyperfinite or is every countable Borel equivalence relation Borel-bounded?

Problem (7.29) Does measure hyperfiniteness imply hyperfiniteness?

Problem (7.32) Let G be a countable amenable group. Is it true that every Borel action of G generates a hyperfinite equivalence relation?

Problem (7.39) If E is Borel hypersmooth (resp., Borel hyperfinite) and T is a Borel function (resp., countable-to-1 Borel function), are the relations $E_0(E, T), E_t(E, T)$ hypersmooth (resp., hyperfinite)?

Problem (7.42) Let E be a Borel equivalence relation such that $E \leq_B E_2$. Is it true that exactly one of the following holds?

(i) E is essentially hyperfinite.
(ii) $E \sim_B E_2$.

Problem (7.46) Let E be a hyperfinite equivalence relation on $\mathbb{N}^{\mathbb{N}}$ that is Δ^1_1 (effectively Borel). Is there a Δ^1_1 automorphism of $\mathbb{N}^{\mathbb{N}}$ such that $E = E_T$? Equivalently, is it true that $E = \bigcup_n E_n$, where (E_n) is a Δ^1_1 (uniformly in n) increasing sequence of finite equivalence relations?

Problem (7.48)

(i) Is there a countable basis for the quasi-order of Borel reducibility $\leq_B$ on the nonhyperfinite Borel equivalence relations?
(ii) Consider the class $\mathcal{B}$ of all equivalence relations of the form E_a, where a is a free Borel action of $\mathbb{F}_2$ admitting an invariant probability measure. Is $\mathcal{B}$ a basis for the quasi-order of Borel reducibility $\leq_B$ on the nonhyperfinite Borel equivalence relations?

Problem (7.50) Consider the class C of countable Borel equivalence relations E that are not μ-hyperfinite for every E-*invariant* probability measure μ, and let C' be the subclass consisting of all equivalence relations of the form E_a, where a is a free Borel action of $\mathbb{F}_2$. Is C' a basis for C for the partial order $\subseteq$ of inclusion?

Problem (7.51) Let E be a countable Borel equivalence relation on a standard Borel space X, and let μ be an E-invariant, E-ergodic probability measure on X. Is it true that exactly one of the following holds?

(i) E is μ-hyperfinite.
(ii) There is an E-invariant Borel set $A \subseteq X$ with $\mu(A) = 1$ and a free Borel action a of $\mathbb{F}_2$ on A such that $E_a \subseteq E \upharpoonright A$.

16.5 Amenability

Problem (8.5) Let E be an amenable countable Borel equivalence relation. Is it true that E is hyperfinite?

Problem (8.7) Let G be an amenable Polish locally compact group. Is it true that any Borel action of G generates an essentially hyperfinite equivalence relation?

Problem (8.12) Is Fréchet amenability equivalent to amenability? Moreover, is Fréchet amenability equivalent to hyperfiniteness?

Problem (8.13) Is the transfinite hierarchy of Fréchet amenability proper, that is, does $\alpha < \beta$ imply that there is a β-amenable Borel equivalence relation that is not α-amenable?

16.6 Treeability

Problem (9.7) Let $E \subseteq F$ be countable Borel equivalence relations such that E is treeable and every F-class contains only finitely many E-classes. Is F treeable?

Problem (9.9)

(i) Is every measure-treeable countable Borel equivalence relation treeable?
(ii) Does the analog of Problem 9.7 have a positive answer in the case of μ-treeability, even for F-invariant μ?

16.7 Contractible Simplicial Complexes

Problem (9.31) If the countable Borel equivalence relation F is C_n-structurable, $n \geq 2$, and $E \sqsubseteq_B F$, is E also C_n-structurable?

16.8 Freeness

Problem (10.4) Let $E \subseteq F$ be countable Borel equivalence relations such that each F-class contains only finitely many E-classes. If E is essentially free, is F also essentially free?

16.9 Universality and Equivalence Relations from Computability Theory

Problem (11.5) Is $\equiv_T$ universal?

Problem (11.7) Is $\cong^2_{\mathrm{rec}}$ universal?

Problem (11.22) Is every weakly universal countable Borel equivalence relation universal?

Problem (11.25) Is it true that for every $f\colon \equiv_T \to_B E_0$, there is a cone of Turing degrees C such that $f(C)$ is contained in a single E_0-class?

Problem (11.31) Is it true that for every universal countable Borel equivalence relation E and every (φ_n) such that $E = R_{(\varphi_n)}$, E is uniformly universal with respect to (φ_n)?

Problem (11.32) Is $\cong^2_{\mathrm{rec}}$ uniformly universal?

16.10 The Poset of Bireducibility Types

Problem (12.3) Is it true that any two bireducibility types have an infimum?

16.11 Structurability

Problem (13.12) Is there a characterization of the $L_{\omega_1\omega}$ theories σ for which every equivalence relation in $\mathcal{E}_\sigma$ is hyperfinite?

Problem (13.13) Is there a characterization of the $L_{\omega_1\omega}$ theories σ for which every aperiodic countable Borel equivalence relation is in $\mathcal{E}_\sigma$?

16.12 Compact Action Realizations

Problem (14.5) Does Turing equivalence $\equiv_T$ have a compact action realization?

Problem (14.6) Does every nonsmooth countable Borel equivalence relation have a compact action realization?

16.13 Complexity of Hyperfiniteness in the Space of Subshifts

Problem (15.2) Is hyperfiniteness comeager in $\mathrm{Sh}(G, \mathbb{I}^{\mathbb{N}})$?

Problem (15.4) What is is the exact descriptive complexity of hyperfiniteness in the space $\mathrm{Sh}(G, \mathbb{I}^{\mathbb{N}})$?

References

[AFP] N. Ackerman, C. Freer and R. Patel. Invariant measures concentrated on countable structures. *Forum Math. Sigma*, **4** (2016), e17, 59pp. 129

[A1] S. Adams. Indecomposability of treed equivalence relations. *Israel J. Math.*, **64**(3) (1988), 362–380. 50, 93, 99

[A2] S. Adams. An equivalence relation that is not freely generated. *Proc. Amer. Math. Soc.*, **102** (1988), 565–566. 95

[A3] S. Adams. Trees and amenable equivalence relations. *Ergodic Theory Dynam. Systems*, **10** (1990), 1–14. 93, 97

[A4] S. Adams. Boundary amenability for word hyperbolic groups and an application to smooth dynamics of simple groups. *Topology*, **33**(4) (1994), 765–783. 73

[A5] S. Adams. Indecomposability of equivalence relations generated by word hyperbolic groups. *Topology*, **33**(4) (1994), 785–798. 99

[A6] S. Adams. Containment does not imply Borel reducibility. In *Set Theory*, DIMACS Ser. Discrete Math. Theoret. Comput. Sci., **58**, American Mathematical Society, 2002, 1–23. 17

[AK] S. Adams and A.S. Kechris. Linear algebraic groups and countable Borel equivalence relations. *J. Amer. Math. Soc.*, **13**(4) (2000), 909–943. 51, 52, 55, 58

[AL] S. Adams and R. Lyons. Amenability, Kazhdan's property and percolation for trees, groups, and equivalence relations. *Israel J. Math.*, **75** (1991), 341–370. 87, 90

[AS] S. Adams and R. Spatzier. Kazhdan groups, cocycles and trees. *Amer. J. Math.*, **112** (1990), 271–287. 100

[Al] S. Allison. Countable Borel treeable equivalence relations are classified by ℓ_1. ArXiv: 2305.01049. 70

[Am] W. Ambrose. Representation of ergodic flows. *Ann. of Math.*, **42** (1941), 723–739. 25

[AP] C. Anantharaman and S. Popa. *An Introduction to* II_1 *Factors*, `math.ucla.edu/~popa/books.html`. xii

[ACH] A. Andretta, R. Camerlo and G. Hjorth. Conjugacy equivalence relation on subgroups. *Fund. Math.*, **167** (2001), 189–212. 112, 113, 118

[B] J. Barwise. *Admissible Sets and Structures*, Springer-Verlag, 1975. 31

[BK] H. Becker and A.S. Kechris. *The Descriptive Set Theory of Polish Group Actions*, Cambridge University Press, 1996. 7, 12, 30, 31, 37

[BY] A. Bernshteyn and J. Yu. Coarse embeddings into grids and asymptotic dimension for Borel graphs of polynomial growth. ArXiv:2302.04737. 70

[BG] S.I. Bezuglyi and V.Ya. Golodets. Hyperfinite and II_1 actions for nonamenable groups. *J. Funct. Anal.*, **40** (1981), 30–44. 91

[Bo] L. Bowen. Finitary random interlacements and the Gaboriau–Lyons problem. *Geom. Funct. Anal.*, **29**(3) (2019), 659–689. 79

[BHI] L. Bowen, D. Hoff and A. Ioana. von Neumann's problem and extensions of non-amenable equivalence relations. *Groups Geom. Dyn.*, **12**(2) (2018), 399–448. 79

[BJ1] C.M. Boykin and S. Jackson. *Some Applications of Regular Markers*. In Lecture Notes in Logic, **24**, Assoc. Symb. Logic, 2006, 138–155. 71

[BJ2] C.M. Boykin and S. Jackson. Borel boundedness and the lattice rounding property. *Contemp. Math.*, **425** (2007), 113–126. 66, 67, 68

[BO] N.P. Brown and N. Ozawa. *C^*-Algebras and Finite-Dimensional Approximations*, Amer. Math. Soc., 2008. 136

[C] P. Calderoni. Rotation equivalence and cocycle superrigidity for compact actions. *J. London Math. Soc.*, **107**(1) (2023), 189–212. 56, 101

[CC] F. Calderoni and A. Clay. Borel structures on the space of left-orderings. *Bull. London Math. Soc.*, **54**(1) (2022), 83–94. 114

[Ca] R. Camerlo. The relation of recursive isomorphism for countable structures. *J. Symb. Logic*, **167**(2) (2002), 879–875. 112

[CG] Y. Carrière and E. Ghys. Relations d'équivalence moyennables sur les groupes de Lie. *C.R. Acad. Sci. Paris Sér. I Math.*, **300**(19) (1985), 677–680. 66, 91

[CM] W. Chan and C. Meehan. Definable combinatorics of some Borel equivalence relations. ArXiv:1709.04567. 77

[CN1] P. Chaube and M.G. Nadkarni. A version of Dye's Theorem for descriptive dynamical systems. *Sankyā, Ser. A*, **49** (1987), 288–304. 12

[CN2] P. Chaube and M.G. Nadkarni. On orbit equivalence of Borel automorphisms. *Proc. Indian Acad. Sci. Math. Sci.*, **99**(3) (1989), 255–261. 12

[Ch1] R. Chen. Borel structurability by locally finite simplicial complexes. *Proc. Amer. Math. Soc.*, **146**(7) (2018), 3085–3096. 105, 106

[Ch2] R. Chen. Decompositions and measures on countable Borel equivalence relations. *Ergodic Theory Dynam. Systems*, **41**(12) (2021), 3671–3703. 39, 106, 125

[Ch3] R. Chen. On uniform ergodic decomposition. `rynchn.github.io/math/`. 39

[CK] R. Chen and A.S. Kechris. Structurable equivalence relations. *Fund. Math.*, **242** (2018), 109–185. See also ArXiv:1606.01995. 14, 85, 88, 95, 106, 126, 128, 129, 130

[CPTT] R. Chen, A. Poulin, R. Tao and A. Tserunyan. Tree-like graphings, wallings, and median graphings of equivalence relations. ArXiv: 2308.13010. 99

[CTT] R. Chen, G. Terlov and A. Tserunyan. Nonamenable subforests of multi-ended quasi-pmp graphs. ArXiv: 2211.07908. 97

[Chr] J.P.R. Christensen. *Topology and Borel Structure*, North-Holland, 1974. 91

[Cl1] J.D. Clemens. Generating equivalence relations by homeomorphisms. `https://citeseerx.ist.psu.edu/pdf/53c95108384476774591de17efc860df49a654bf`, 2008. 6

[Cl2] J.D. Clemens. Isomorphism of subshifts is a universal countable Borel equivalence relation. *Israel J. Math.*, **170** (2009), 113–123. 5, 73, 114

[Cl3] J.D. Clemens. Isomorphism and weak conjugacy of free Bernoulli subflows. *Contemp. Math.*, **2020**, 77–87. 114

[CCM] J.D. Clemens, C. Conley and B.D. Miller. Borel homomorphisms of smooth σ-ideals. `https://citeseerx.ist.psu.edu/document?repid=rep1&type=pdf&doi=2255f1001874e0ab9943a964ee6c3a3fecfc2b15`, 2007. 128

[CLM] J.D. Clemens, D. Lecomte and B.D. Miller. Essential countability of treeable equivalence relations. *Adv. Math.*, **265** (2014), 1–31. 29

[CGMT] C.T. Conley, D. Gaboriau, A.S. Marks and R.D. Tucker-Drob. One-ended spanning subforests and treeability of groups. ArXiv:2104.07431. 90, 98, 99

[CJMST1] C.T. Conley, S. Jackson, A.S. Marks, B. Seward and R.D. Tucker-Drob. Hyperfiniteness and Borel combinatorics. *J. Eur. Math. Soc.*, **22**(3) (2020), 877–892. 77

[CJMST2] C.T. Conley, S. Jackson, A.S. Marks, B. Seward and R.D. Tucker-Drob. Borel asymptotic dimension and hyperfinite equivalence relations. *Duke Math. J.*, to appear, ArXiv:2009.06721. 71

[CKM] C.T. Conley, A.S. Kechris and B.D. Miller. Stationary probability measures and topological realizations. *Israel J. Math.*, **198** (2013), 333–345. 14, 37

[CMa] C.T. Conley and A.S. Marks. Distance from marker sequences in locally finite Borel graphs. *Contemp. Math.*, **752** (2020), 89–92. 11

[CM1] C.T. Conley and B.D. Miller. Measure reducibility of countable Borel equivalence relations. *Ann. of Math.*, **185**(2) (2017), 347–402. 9, 16, 17, 51, 54, 64, 73, 78, 103, 104

[CM2] C.T. Conley and B.D. Miller. Incomparable actions of free groups. *Ergodic Theory Dynam. Systems*, **37** (2017), 2084–2098. 54, 103, 105

[CFW] A. Connes, J. Feldman and B. Weiss. An amenable equivalence relation is generated by a single transformation. *Ergodic Theory Dynam. Systems*, **1** (1981), 431–450. 72, 89, 90, 91

[Co1] S. Coskey. Descriptive aspects of torsion-free abelian groups. *Ph.D. Thesis*, Rutgers University, 2008. `scoskey.org/publications`. 58

[Co2] S. Coskey. Borel reductions of profinite actions of $\mathrm{SL}_n(\mathbb{Z})$. *Ann. Pure Appl. Logic*, **161** (2010), 1270–1279. 56, 58

[Co3] S. Coskey. The classification of torsion-free abelian groups of finite rank up to isomorphism and up to quasi-isomorphism. *Trans. Amer. Math. Soc.*, **364**(1) (2012), 175-194. 58

[Co4] S. Coskey. Ioana's superrigidity theorem and orbit equivalence relations. *ISRN Algebra*, Art. ID 387540, 8pp. 56, 58

[CS] S. Coskey and S. Schneider. Cardinal characteristics and countable Borel equivalence relations. *Math. Logic Q.*, **63**(3–4) (2017), 211–227. 67

[Cot] M. Cotton. Abelian group actions and hypersmooth equivalence relations. *Ann. Pure Appl. Logic*, **173(8)** (2022), 103–122. 70

[DaMa] A.R. Day and A.S. Marks. On a question of Slaman and Steel. ArXiv: 2004.00174. 80

[DM] C. Dellacherie and P.-A. Meyer. *Théorie Discrète du Potentiel*, Hermann, 1983. 91

[dRM1] N. de Rancourt and B.D. Miller. The Feldman–Moore, Glimm–Effros, and Lusin–Novikov theorems over quotients. *J. Symb. Logic*, to appear. ArXiv:2105.05374. 20

[dRM2] N. de Rancourt and B.D. Miller. A dichotomy for countable unions of smooth equivalence relations. *J. Symb. Logic*, to appear. ArXiv:2105.05362. 20, 33

[DG] L. Ding and S. Gao. Non-archimedean Polish groups and their actions. *Adv. Math.*, **307** (2017), 312–343. 70

[Di] A. Ditzen. Definable Equivalence Relations on Polish Spaces. *Ph.D. Thesis*, Caltech, 1992. 37, 39, 41, 44, 45

[DJK] R. Dougherty, S. Jackson and A.S. Kechris. The structure of hyperfinite Borel equivalence relations. *Trans. Amer. Math. Soc.*, **341**(1) (1994), 193–225. xi, 12, 13, 15, 16, 36, 40, 41, 49, 59, 60, 61, 62, 63, 66, 72, 77, 107, 108

[DK] R. Dougherty and A.S. Kechris. How many Turing degrees are there? *Contemp. Math.*, **257** (2000), 83–94. 112

[Dy] H.A. Dye. On groups of measure preserving transformations, I. *Amer. J. Math.*, **81** (1959), 119–159; II. *Amer. J. Math.*, **85** (1963), 551–576. 61

[E1] E.G. Effros. Transformation groups and C^*-algebras. *Ann. of Math.*, **81**(2) (1965), 38–55. 36, 40, 48

[E2] E.G. Effros. Polish transformation groups and classification problems. In *General Topology and Modern Analysis*, Academic Press, 1980, 217–227. 36, 40, 48

[EOSS] S.K. Elayavalli, K. Oyakawa, F. Shinko and P. Spaas. Hyperfiniteness for group actions on trees, ArXiv:2307.10964. 73

[El] G. Elek. Finite graphs and amenability. *J. Funct. Anal.*, **263** (2012), 2593–2614 89, 90

[EH] I. Epstein and G. Hjorth. Rigidity and equivalence relations with infinitely many ends. `logic.ucla.edu/greg/research.html`. 83

[ET] I. Epstein and T. Tsankov. Modular actions and amenable representations. *Trans. Amer. Math. Soc.*, **362**(2) (2010), 603–621. 54, 102, 103

[Fa] R.H. Farrell. Representation of invariant measures. *Ill. J. Math.*, **6** (1962), 447–467. 38

[FHM] J. Feldman, P. Hahn and C.C. Moore. Orbit structure and countable sections for actions of continuous groups. *Adv. Math.*, **26** (1979), 186–230. 25, 28

[FM] J. Feldman and C.C. Moore. Ergodic equivalence relations and von Neumann algebras, I. *Trans. Amer. Math. Soc.*, **234** (1977), 289–324. 5, 68

[Fo] P. Forrest. On the virtual groups defined by ergodic actions of $\mathbb{R}^n$ and $\mathbb{Z}^n$. *Adv. Math.*, **14** (1974), 271–308. 25

[FKS] J. Frisch, A.S. Kechris and F. Shinko. Lifts of Borel actions on quotient spaces. *Israel J. Math.*, **251** (2022), 379–421. 20, 21, 22

[FKSV] J.R. Frisch, A.S. Kechris, F. Shinko and Z. Vidnyánszky. Realizations of countable Borel equivalence relations. ArXiv:2109.12486. 14, 63, 64, 71, 117, 121, 131, 132, 133, 134, 135, 136

[FS] J. Frisch and F. Shinko. Quotients by countable subgroups are hyperfinite. *Groups Geom. Dyn.*, **17**(3) (2023), 985–992. 74

[Fu] A. Furman. Orbit equivalence rigidity. *Ann. of Math.*, **150** (1999), 1083–1108. 95, 109

[Ga1] D. Gaboriau. Coût des relations d'équivalence et des groupes. *Invent. Math.*, **139** (2000), 41–98. 6, 93, 97, 100

[Ga2] D. Gaboriau. Invariantes ℓ^2 de relations d'équivalence et des groupes. *Publ. Math. Inst. Hautes Études Sci.*, **95** (2002), 93–150. 106

[GL] D. Gaboriau and R. Lyons. A measurable-group-theoretic solution to von Neumann's problem. *Invent. Math.*, **177**(3) (2009), 533–540. 79

[G1] S. Gao. The action of $\mathrm{SL}_2(\mathbb{Z})$ on the subsets of $\mathbb{Z}^2$. *Proc. Amer. Math. Soc.*, **129**(5) (2000), 1507–1512. 114

[G2] S. Gao. Coding subset shift by subgroup conjugacy. *Bull. London Math. Soc.*, **132**(6) (2000), 1653–1657. 113

[G3] S. Gao. Some applications of the Adams–Kechris technique. *Proc. Amer. Math. Soc.*, **130**(3) (2002), 863–874. 52

[GH] S. Gao and A. Hill. Topological isomorphism for rank-1 systems. *J. Anal. Math.*, **128**(1) (2016), 1–49. 74

[GJ] S. Gao and S. Jackson. Countable abelian groups actions and hyperfinite equivalence relations. *Invent. Math.*, **201** (2015), 309–383. 70, 71

[GJKS] S. Gao, S. Jackson, E. Krohne and B. Seward. Forcing constructions and countable Borel equivalence relations. *J. Symb. Logic*, **87**(3) (2022), 873–893. 11

[GJS1] S. Gao, S. Jackson and B. Seward. A coloring property for countable groups. *Math. Proc. Cambridge Phil. Soc.*, **147**(3) (2009), 579–592. 11

[GJS2] S. Gao, S. Jackson and B. Seward. Group colorings and Bernoulli subflows. *Mem. Amer. Math. Soc.*, **241**(1141) (2016). 74, 114

[GK] S. Gao and A.S. Kechris. On the classification of Polish metric spaces up to isometry. *Mem. Amer. Math. Soc.*, **161**(766) (2016). 32, 74, 115

[GdlH] T. Giordano and P. de la Harpe. Moyennabilité des groupes dénombrables et actions sur les espaces de Cantor. *C.R. Acad. Sci. Paris Sér. I Math.*, **324**(11) (1997), 1255–1258. 134

[GPS] T. Giordano, I. Putnam and C. Skau. Affable equivalent relations and orbit structure of Cantor dynamical systems. *Ergodic Theory Dynam. Systems*, **23** (2004), 441–475. 6

[GW] E. Glasner and B. Weiss. On the interplay between measurable and topological dynamics. In *Handbook of Dynamical Systems*, Vol. 1B. B. Hasselblatt and A. Katok (eds). Elsevier, 2006, 597–648. 61

[Gr1] J. Grebik. σ-lacunary actions of Polish groups. *Proc. Amer. Math. Soc.*, **148**(8) (2020), 3583–3589. 27, 33, 70

[Gr2] J. Grebik. Borel equivalence relations induced by actions of tsi Polish groups. ArXiv:2107.14439. 33

[HaK] A.B. Hajian and S. Kakutani. Weakly wandering sets and invariant measures. *Trans. Amer. Math. Soc.*, **110** (1964), 136–151. 41

[HMT] A. Halbäck, M. Malicki and T. Tsankov. Continuous logic and Borel equivalence relations. *J. Symb. Logic*, **88**(4), 1725–1752. 25

[HKL] L.A. Harrington, A.S. Kechris and A. Louveau. A Glimm–Effros dichotomy for Borel equivalence relations. *J. Amer. Math. Soc.*, **3**(4) (1990), 903–928. 30, 36, 48

[HL] K. Higuchi and P. Lutz. A note on a question of Sacks: It is harder to embed height-three partial orders than to embed height-two partial orders. ArXiv: 2309.01876. 116

[H1] G. Hjorth. Around nonclassifiability for countable torsion free abelian groups. In *Abelian Groups and Modules, Dublin, 1998*, Birkhauser, 1999, 269–292. 57, 100

[H2] G. Hjorth. Actions by the classical Banach spaces. *J. Symb. Logic*, **65**(1) (2000), 392–420. 33

[H3] G. Hjorth. A converse to Dye's theorem. *Trans. Amer. Math. Soc.*, **357**(8) (2005), 3083–3103. 54, 101, 102

[H4] G. Hjorth. Bi-Borel reducibility of essentially countable Borel equivalence relations. *J. Symb. Logic*, **70**(3) (2005), 979–992. 23, 96

[H5] G. Hjorth. A dichotomy for being essentially countable. *Contemp. Math.*, **380** (2005), 109–127. 24, 25

[H6] G. Hjorth. The Furstenberg lemma characterizes amenability. *Proc. Amer. Math. Soc.*, **134**(10) (2006), 3061–3069. 90

[H7] G. Hjorth. A lemma for cost attained. *Ann. Pure Appl. Logic*, **143**(1–3) (2006), 87–102. 95

[H8] G. Hjorth. Borel equivalence relations which are highly unfree. *J. Symb. Logic*, **73**(4) (2008), 1271–1277. 109

[H9] G. Hjorth. Non-treeability for product group actions. *Israel J. Math.*, **163** (2008), 383–409. 100

[H10] G. Hjorth. Selection theorems and treeability. *Proc. Amer. Math. Soc.*, **136**(10) (2008), 3647–3653. 29

[H11] G. Hjorth. Two generated groups are universal. `logic.ucla.edu/greg/research.html`. 113

[H12] G. Hjorth. Treeable equivalence relations. *J. Math. Logic*, **12**(1) (2012), 1250003, 21pp. 17, 54, 103

[HK1] G. Hjorth and A.S. Kechris. Borel equivalence relations and classifications of countable models. *Ann. Pure Appl. Logic*, **82** (1996), 221–272. 30, 31, 40, 65, 94, 98, 100

[HK2] G. Hjorth and A.S. Kechris. The complexity of the classification of Riemann surfaces and complex manifolds. *Ill. J. Math.*, **44**(1) (2000), 104–137. 26, 82, 83, 114

[HK3] G. Hjorth and A.S. Kechris. Recent developments in the theory of Borel reducibility. *Fund. Math.*, **170** (2001), 21–52. 33, 75

[HK4] G. Hjorth and A.S. Kechris. Rigidity theorems for actions of product groups and countable Borel equivalence relations. *Mem. Amer. Math. Soc.*, **833** (2005). 16, 17, 51, 54, 57, 58, 101, 106

[HKLo] G. Hjorth, A.S. Kechris and A. Louveau. Borel equivalence relations induced by actions of the symmetric group. *Ann. Pure Appl. Logic*, **92** (1998), 63–112. 30

[Ho] M. Hochman. Every Borel automorphism without finite invariant measure admits a two-set generator. *J. Eur. Math. Soc.*, **21**(1) (2019), 271–317. 117

[HJ] J. Holshouser and S. Jackson. Partition properties for hyperfinite quotients. *Preprint*. 77

[HSS] S.J. Huang, M. Sabok and F. Shinko. Hyperfiniteness of boundary actions of cubulated hyperbolic groups. *Ergodic Theory Dynam. Systems*, **40**(9) (2020), 2453–2466. 73

[I1] A. Ioana. Relative property (T) for the subequivalence relations induced by the action of $\mathrm{SL}_2(\mathbb{Z})$ on $\mathbb{T}^2$. *Adv. Math.*, **224**(4) (2010), 1589–1617. 103

[I2] A. Ioana. Orbit equivalence and Borel reducibility rigidity for profinite actions with spectral gap. *J. Eur. Math. Soc.*, **18**(12) (2016), 2733–2784. 56, 96, 105

[I3] A. Ioana. Rigidity for von Neumann algebras. In *Proc. of ICM, Rio de Janeiro, 2018, Vol. III: Invited Lectures*, World Scientific (2018), 1639–1672. xii

[I4] A. Ioana. Compact actions whose orbit equivalence relations are not profinite. *Adv. Math.*, **354** (2019), 106753, 19pp. 103

[IS] S. Iyer and F. Shinko. The generic action of the free group on Cantor space. *In preparation*. 136

[JKL] S. Jackson, A.S. Kechris and A. Louveau. Countable Borel equivalence relations. *J. Math. Logic*, **2**(1) (2002), 1–80. xi, 6, 50, 62, 66, 69, 70, 73, 81, 82, 83, 84, 86, 87, 90, 91, 94, 95, 97, 98, 99, 101, 107, 108, 110, 113, 116, 117, 118

[Kai] V.A. Kaimanovich. Amenability, hyperfiniteness, and isoperimetric inequalities, *C. R. Acad. Sci. Paris Ser. I Math.*, **325**(9) (1997), 999–1004. 81, 89, 90

[Ka] V. Kanovei. *Borel Equivalence Relations*, American Mathematical Society, 2008. 24, 33

[KSZ] V. Kanovei, M. Sabok and J. Zapletal. *Canonical Ramsey Theory on Polish Spaces*, Cambridge University Press, 2013. 34

[KW] Y. Katznelson and B. Weiss. The classification of non-singular actions, revisited. *Ergodic Theory Dynam. Systems*, **11**(2) (1991), 333–348. 62

[Kay] B. Kaya. The complexity of the topological conjugacy problem for Toeplitz subshifts. *Israel J. Math.*, **220** (2017), 873–897. 74

[Ke1] A.S. Kechris. Amenable equivalence relations and Turing degrees. *J. Symb. Logic*, **56**(1) (1991), 182–194. 87

[Ke2] A.S. Kechris. Countable sections for locally compact group actions. *Ergodic Theory Dynam. Systems*, **12** (1992), 283–295. 24, 25, 28, 29

[Ke3] A.S. Kechris. Amenable versus hyperfinite Borel equivalence relations. *J. Symb. Logic*, **58**(3) (1993), 894–907. 91

[Ke4] A.S. Kechris. Countable sections for locally compact group actions, II. *Proc. Amer. Math. Soc.*, **120**(1) (1994), 241–247. 26, 62

[Ke5] A.S. Kechris. Lectures on definable group actions and equivalence relations. *Preprint*, 1994. 37, 39, 41, 42, 45, 64, 65, 68

[Ke6] A.S. Kechris. *Classical Descriptive Set Theory*, Springer, 1995. 5, 25, 28, 31, 38, 42, 77

[Ke7] A.S. Kechris. On the classification problem for rank 2 torsion-free abelian groups. *J. London Math. Soc.*, **62**(2) (2000), 437–450. 100, 101

[Ke8] A.S. Kechris. Descriptive dynamics. In *Descriptive Set Theory and Dynamical Systems*, London Mathematical Society Lecture Note Series, **277**, Cambridge University Press, 2000, 231–258. 25

[Ke9] A.S. Kechris. Unitary representations and modular actions. *J. Math. Sci.*, **140**(3) (2007), 398–425. 54, 101, 102

[Ke10] A.S. Kechris. *Global Aspects of Ergodic Group Actions*, American Mathematical Society, 2010. 17, 54, 109

[Ke11] A.S. Kechris. The spaces of measure preserving equivalence relations and graphs. `pma.caltech.edu/people/alexander-kechris`. 5, 106

[Ke12] A.S. Kechris. Quasi-invariant measures for continuous group actions. *Contemp. Math.*, **752** (2020), 113–120. 42

[KL] A.S. Kechris and A. Louveau. The classification of hypersmooth Borel equivalence relations. *J. Amer. Math. Soc.*, **10**(1) (1997), 215–242. 75

[KMa] A.S. Kechris and H.L. Macdonald. Borel equivalence relations and cardinal algebras. *Fund. Math.*, **235** (2016), 183–198. 24, 124, 125

[KMPZ] A.S. Kechris, M. Malicki, A. Panagiotopoulos and Z. Zielinski. On Polish groups admitting non-essentially countable actions. *Ergodic Theory Dynam. Systems*, **42** (2022), 180–194. 28

[KM] A.S. Kechris and A.S. Marks. Descriptive graph combinatorics. `pma.caltech.edu/people/alexander-kechris`. xii

[KM1] A.S. Kechris and B.D. Miller. *Topics in Orbit Equivalence*, Springer, 2004. xii, 11, 13, 14, 38, 43, 44, 61, 65, 79, 81, 93, 94, 99

[KM2] A.S. Kechris and B.D. Miller. Means on equivalence relations. *Israel J. Math.*, **163** (2008), 241–262. 9

[KST] A.S. Kechris, S. Solecki and S. Todorcevic. Borel chromatic numbers. *Adv. Math.*, **141** (1999), 1–44. 5, 63, 126

[KWo] A.S. Kechris and M. Wolman. Ditzen's effective version of Nadkarni's Theorem. `pma.caltech.edu/people/alexander-kechris`. 37, 39

[KeL] D. Kerr and H. Li. *Ergodic Theory*, Springer, 2016. 90

[Kh] A. Khezeli. Shift-coupling of random rooted graphs and networks. *Contemp. Math.*, **719** (2018), 175–211. 45

[KT] Y. Kida and R. Tucker-Drob. Inner amenable groupoids and central sequences. *Forum Math. Sigma*, **8** (2020), Paper No. e29, 84 pp. 90

[Kr] W. Krieger. On non-singular transformations of a measure space, I. *Z. Wahrsch. Verw. Gebiete*, **11** (1969), 83–97; II. *ibid*, **11** (1969), 98–119. 62

[KPS] K. Krupinski, A. Pillay and S. Solecki. Borel equivalence relations and Lascar strong types. *J. Math. Logic*, **13**(2) (2013), 1350008, 37 pp. 74

[KPV] D. Kyed, H.D. Petersen and S. Vaes. L^2-Betti numbers of locally compact groups and their cross section equivalence relations. *Trans. Amer. Math. Soc.*, **367**(7) (2015), 4917–4956. 26, 37

[La] P.B. Larson. The filter dichotomy and medial limits. *J. Math. Logic*, **9**(2) (2009), 159–165. 91

[Lec] D. Lecomte. On the complexity of Borel equivalence relations with some countability property. *Trans. Amer. Math. Soc.*, **373**(3), (2020), 1845–1883. 22

[Le] G. Levitt. On the cost of generating an equivalence relation. *Ergodic Theory Dynam. Systems*, **15** (1995), 1173–1181. 93

[LE] E.G.K. Lopez-Escobar. An interpolation theorem for denumerably long formulas. *Fund. Math.*, **57** (1965), 253–272. 127

[LM] A. Louveau and G. Mokobodzki. On measures ergodic with respect to an analytic equivalence relation. *Trans. Amer. Math. Soc.*, **349**(12) (1997), 4815–4823. 45

[L] M. Lupini. Polish groupoids and functorial complexity. *Trans. Amer. Math. Soc.*, **369**(9) (2017), 6683–6723. 95

[LS] P. Lutz and B. Siskind. Part 1 of Martin's conjecture for order preserving and measure preserving functions. ArXiv: 2305.19646. 116

[Ma] M. Malicki. Abelian pro-countable groups and orbit equivalence relations. *Fund. Math.*, **233**(1) (2016), 83–99. 28

[Mar] D. Marker. The Borel complexity of isomorphism for theories with many types. *Notre Dame J. Formal Logic*, **48**(1) (2007), 93–97. 32

[M1] A.S. Marks. A determinacy approach to Borel combinatorics. *J. Amer. Math. Soc.*, **29**(2) (2016), 579–600. 10

[M2] A.S. Marks. The universality of polynomial time Turing equivalence. *Math. Structures Comput. Sci.*, **28**(3) (2018), 448–456. 112

[M3] A.S. Marks. Uniformity, universality, and computability theory. *J. Math. Logic*, **17** (2017), 1750003, 50 pp. 11, 111, 112, 119, 120, 121, 126, 127, 129

[M4] A.S. Marks. A short proof of the Connes-Feldman–Weiss theorem. `math.berkeley.edu/~marks/`. 68, 78, 90

[MSS] A.S. Marks, T.A. Slaman and J.R. Steel. *Martin's Conjecture, Arithmetic Equivalence and Countable Borel Equivalence Relations*. In Lecture Notes in Logic, Assoc. Symb. Logic, **43**, 2016, 493–519. 110, 111, 119

[Marq] J. Marquis. On geodesic ray bundles in buildings. *Geom. Dedicata*, **202** (2019), 27–43. 73

[MS] J. Marquis and M. Sabok. Hyperfiniteness of boundary actions of hyperbolic groups. *Math. Ann.*, **377**(3–4) (2020), 1129–1153. 73

[Me] R. Mercer. The full group of a countable measurable equivalence relation. *Proc. Amer. Math. Soc.*, **117**(2) (1993), 323–333. 18

[Mi1] B.D. Miller. A bireducibility lemma. Item 4 under Unpublished in `glimmeffros.github.io`. 15

[Mi2] B.D. Miller. Borel dynamics. *preprint*, 2007. 18, 44

[Mi3] B.D. Miller. A generalized marker lemma. Item 2 under Unpublished in `glimmeffros.github.io`. 9

[Mi4] B.D. Miller. The classification of finite Borel equivalence relations on $2^{\mathbb{N}}/E_0$. Item 12 under Unpublished in `glimmeffros.github.io`. 76

[Mi5] B.D. Miller, Full groups, classification, and equivalence relations. *Ph.D. Thesis*, U.C. Berkeley, 2004. Item 2 under Publications in `glimmeffros.github.io`. 16, 18, 20

[Mi6] B.D. Miller, Borel equivalence relations and everywhere faithful actions of free products. Item 8 under Unpublished in `glimmeffros.github.io`. 109

[Mi7] B.D. Miller. On the existence of invariant probability measures for Borel actions of countable semigroups. Item 9 under Unpublished in `glimmeffros.github.io`. 14

[Mi8] B.D. Miller. The existence of measures of a given cocycle, I: atomless, ergodic σ-finite measures. *Ergodic Theory Dynam. Systems*, **28** (2008), 1599–1613. 44

[Mi9] B.D. Miller. The existence of measures of a given cocycle, II: probability measures. *Ergodic Theory Dynam. Systems*, **28** (2008), 1615–1633. 44, 45

[Mi10] B.D. Miller. Ends of graphed equivalence relations, I. *Israel J. Math.*, **169** (2009), 375–392 74

[Mi11] B.D. Miller. Incomparable treeable equivalence relations. *J. Math. Logic*, **12**(1) (2012), 1250004, 11pp. 17, 54, 103

[Mi12] B.D. Miller. On the existence of cocycle-invariant probability measures. *Ergodic Theory Dynam. Systems*, **40**(11) (2020), 3150–3168. 44

[Mi13] B.D. Miller. A generalization of the $\mathbb{G}_0$-dichotomy and a strengthening of the $\mathbb{E}_0^{\mathbb{N}}$-dichotomy. *J. Math. Logic*, **22**(1) (2022), Paper No. 2150028, 19pp. 27, 33

[Mi14] B.D. Miller. The existence of invariant measures. Item 3 under Seminars in `glimmeffros.github.io`. 40, 44

[Mi15] B.D. Miller. Reducibility of countable equivalence relations. Item 4 under Seminars in `glimmeffros.github.io`. 103

[Mi16] B.D. Miller. Compositions of periodic automorphisms. Item 2 under Recent in `glimmeffros.github.io`. 18

[Mi17] B.D. Miller. Essential values of cocycles and the Borel structure of $\mathbb{R}/\mathbb{Q}$. Item 3 under Recent in `glimmeffros.github.io`. 76

[Mi18] B.D. Miller. A first-order characterization of the existence of invariant measures. Item 4 under Recent in `glimmeffros.github.io`. 18

[Mi19] B.D. Miller. A generalization of the Dye–Krieger theorem. Item 23 under Unpublished in `glimmeffros.github.io`. 68

[MR1] B.D. Miller and C. Rosendal. Isomorphism of Borel full groups. *Proc. Amer. Math. Soc.*, **135**(2) (2007), 517–522. 17, 18

[MR2] B.D. Miller and C. Rosendal. Descriptive Kakutani equivalence. *J. Eur. Math. Soc.*, **12**(1) (2010), 179–219. 62

[Mo] J.T. Moore. A brief introduction to amenable equivalence relations. *Contemp. Math.*, **752** (2020), 153–164. 66, 91, 92

[My] J. Mycielski. (untitled) *Amer. Math. Monthly*, **82**(3) (1975), 308–309. 72

[N1] M.G. Nadkarni. Descriptive ergodic theory. *Contemp. Math.*, **94** (1989), 191–209. 12, 13

[N2] M.G. Nadkarni. On the existence of a finite invariant measure. *Proc. Indian Acad. Sci. Math. Sci*, **100** (1991), 203–220. 12, 37

[N3] M.G. Nadkarni. *Basic Ergodic Theory*, 3rd edition, Birkhäuser, 2013. 41

[NV] P. Naryshkin and A. Vaccaro. Hyperfiniteness and Borel asymptotic dimension of boundary actions of hyperbolic groups. ArXiv: 2306, 02056. 73

[Ne] C. Nebbia. Amenability and Kunze–Stein property for groups acting on a tree. *Pacific J. Math.*, **135** (1988), 371–380. 87

[OW] D. Ornstein and B. Weiss. Ergodic theory and amenable group actions, I: The Rohlin lemma. *Bull. Amer. Math. Soc. (NS)*, **2** (1980), 161–164. 69

[O] K. Oyakawa. Hyperfiniteness of boundary actions of acylindrically hyperbolic groups. ArXiv: 2307.09790. 73

[PW] A. Panagiotopoulos and A. Wang. Every CBER is smooth below the Carlson–Simpson generic partition. ArXiv:2206.14224. 34

[Pa] A.L.T. Paterson. *Amenability*, American Mathematical Society, 1988. 81

[Pi] O. Pikhurko. Borel combinatorics of locally finite graphs. In *Surveys in Combinatorics*, London Mathematical Society Lecture Note Series, **470**, 2021, 267–319. xii

[Po] S. Popa. Cocycle and orbit equivalence superrigidity for malleable actions of w-rigid groups. *Invent. Math.*, **170**(2) (2007), 243–295. 54

[PS] P. Przytycki and M. Sabok. Unicorn paths and hyperfiniteness for the mapping class group. *Forum Math. Sigma*, **9** (2021), Paper No. e36, 10pp. 73

[Ro] C. Rosendal. On the non-existence of certain group topologies. *Fund. Math.*, **183**(3) (2005), 213–228. 18

[Q] V. Quorning. Superrigidity, Profinite Actions and Borel Reducibility. *Master's Thesis*, University of Copenhagen, 2015. 58

[R] A. Ramsey. Topologies on measure groupoids. *J. Funct. Anal.*, **47** (1982), 314–343. 28

[Ru] D. Rudolph. A two-valued step coding for ergodic flows. *Math. Z.*, **150**(3) (1976), 201–220. 27

[ST] M. Sabok and T. Tsankov. On the complexity of topological conjugacy of Toeplitz subshifts. *Israel J. Math.*, **220** (2017), 583–603. 74

[Sc1] S. Schneider. Simultaneous reducibility of pairs of Borel equivalence relations. ArXiv:1310.8028. 48

[ScSe] S. Schneider and B. Seward. Locally nilpotent groups and hyperfinite equivalence relations. ArXiv: 1308.5853. 71

[Se] M. Segal. Hyperfiniteness. *Ph.D. Thesis*, Hebrew University of Jerusalem, 1997. 77, 78

[SeT] B. Seward and R.D. Tucker-Drob. Borel structurability of the 2-shift of a countable group. *Ann. Pure Appl. Logic*, **167**(1) (2016), 1–21. 98, 107

[Sh] F. Shinko. Equidecomposition in cardinal algebras. *Fund. Math.*, **253**(2) (2021), 197–204. 46

[Si] J.H. Silver. Counting the number of equivalence classes of Borel and coanalytic equivalence relations. *Ann. Math. Logic*, **18** (1980), 1–28. 47

[SlSt] T.A. Slaman and J.R. Steel. Definable functions on degrees. In *Cabal Seminar 81–85*, Lecture Notes in Mathematics, **1333**, 1988, Springer-Verlag, 37–55. 9, 50, 59, 80, 101

[Sl1] K. Slutsky. Lebesgue orbit equivalence of multidimensional Borel flows: a picturebook of tilings. *Ergodic Theory Dynam. Systems*, **37**(6) (2017), 1966–1996. 26, 27, 37, 62

[Sl2] K. Slutsky. Regular cross sections of Borel flows. *J. Eur. Math. Soc.*, **21**(7) (2019), 1985–2050. 13, 27, 62

[Sl3] K. Slutsky. Cross sections of Borel flows with restrictions on the distance set. ArXiv: 1604.02215. 27

[Sl4] K. Slutsky. On time change equivalence of Borel flows. *Fund. Math.*, **247**(1) (2019), 1–24. 27, 62

[Sl5] K. Slutsky. Countable Borel equivalence relations. *Preprint*, `kslutsky.com`. 13, 37, 39, 60, 61

[Sm] I.B. Smythe. Equivalence of generics. *Arch. Math. Logic*, **61**(5–6) (2022), 795–812. 66

[So] S. Solecki. Actions of non-compact and non-locally compact groups. *J. Symb. Logic*, **65**(4) (2000), 1881–1894. 27, 28

[SWW] D. Sullivan, B. Weiss and J.D.M. Wright. Generic dynamics and monotone complete C^*-algebras. *Trans. Amer. Math. Soc.*, **295**(20) (1986), 795–809. 65

[Ta] A. Tarski. *Cardinal Algebras*, Oxford University Press, 1949. 123, 124, 125

[T1] S. Thomas. The classification problem for p-local torsion-free abelian groups of finite rank. Preprint, `sites.math.rutgers.edu/~sthomas/papers.html`. 56

[T2] S. Thomas. Some applications of superrigidity to Borel equivalence relations. In *DIMACS Ser. Discrete Math. Theoret. Comput. Sci.*, **58** (2002), American Mathematical Society, 129–134. 16

[T3] S. Thomas. The classification problem for torsion-free abelian groups of finite rank. *J. Amer. Math. Soc.*, **16**(1) (2003), 233–258. 57, 58, 98, 100

[T4] S. Thomas. Superrigidity and countable Borel equivalence relations. *Ann. Pure Appl. Logic*, **120** (2003), 237–262. 56, 58

[T5] S. Thomas. Borel superrigidity and the classification problem for the torsion-free abelian groups of finite rank. In *Proc. ICM*, European Mathematical Society, 2006, 93–116. 57, 58

[T6] S. Thomas. Property (τ) and countable Borel equivalence relations. *J. Math. Logic*, **7**(1) (2007), 1–34. 56

[T7] S. Thomas. The classification problem for finite rank Butler groups. In *Models, Modules and Abelian Groups*, Walter de Gruyter, 2008, 329–338. 73

[T8] S. Thomas. On the complexity of the quasi-isometry and virtual isomorphism problems for finitely generated groups. *Groups Geom. Dyn.*, **2** (2008), 281–307. 32, 115

[T9] S. Thomas. Continuous versus Borel reductions. *Arch. Math. Logic*, **48** (2009), 761–770. 71

[T10] S. Thomas. The commensurability relation for finitely generated groups. *J. Group Theory*, **12** (2009), 901–909. 113

[T11] S. Thomas. Martin's conjecture and strong ergodicity. *Arch. Math. Logic*, **48** (2009), 749–759. 67, 118, 119

[T12] S. Thomas. Popa superrigidity and countable Borel equivalence relations. *Ann. Pure Appl. Logic*, **158** (2009), 175–189. 16, 17, 51, 108, 109, 110, 119

[T13] S. Thomas. The classification problem for S-local torsion-free abelian groups of finite rank. *Adv. Math.*, **226** (2011), 3699–3723. 58

[T14] S. Thomas. Universal Borel actions of countable groups. *Groups Geom. Dyn.*, **6** (2012), 389–407. 107, 109, 113, 116, 117, 119

[T15] S. Thomas. On the E_0-extensions of countable Borel equivalence relations. *Preprint*, `sites.math.rutgers.edu/~sthomas/papers.html`. 49

[TS] S. Thomas and S. Schneider. Countable Borel equivalence relations. In *Appalachian Set Theory, 2006–2012*, London Mathematical Society Lecture Note Series, **406**, 2013, 25–62. 58

[TV1] S. Thomas and B. Velickovic. On the complexity of the isomorphism relation for finitely generated groups. *J. Algebra*, **217** (1999), 352–373. 113

[TV2] S. Thomas and B. Velickovic. On the complexity of the isomorphism relation for fields of finite transcendence degree. *J. Pure Appl. Alg.*, **159** (2001), 347–363. 113

[TW] S. Thomas and J. Williams. The bi-embeddability relation for finitely generated groups, II. *Arch. Math. Logic*, **55** (2016), 483–500. 118

[Tho] A. Thompson. A metamathematical condition equivalent to the existence of a complete left invariant metric. *J. Symb. Logic*, **71**(4) (2006), 1108–1124. 27

[Th] H. Thorisson. Transforming random elements and shifting random fields. *Ann. Probab.*, **24**(4) (1996), 2057–2064. 45

[Ts1] A. Tserunyan. Hjorth's proof of the embeddability of hyperfinite equivalence relations into E_0. Preprint, `math.mcgill.ca/atserunyan/research.html`. 60

[Ts2] A. Tserunyan. Segal's effective witness to measure-hyperfiniteness. Preprint, `math.mcgill.ca/atserunyan/research.html`. 68, 78

[Ts3] A. Tserunyan. Finite Generators for Countable Group Actions; Finite Index Pairs of Equivalence Relations; Complexity Measures for Recursive Programs. *Ph.D. Thesis*, UCLA, 2013. `math.mcgill.ca/atserunyan/research.html`. 96

[Ts4] A. Tserunyan. Finite generators for countable group actions in the Borel and Baire category settings. *Adv. Math.*, **269** (2015), 585–646. 117

[Ts5] A. Tserunyan. A descriptive construction of trees and Stallings' theorem. *Contemp. Math.*, **752**, 191–207. 99

[TT] A. Tserunyan and R. Tucker-Drob. The Radon–Nikodym topography of amenable equivalence relations in an acyclic graph. *In preparation*. 97

[Va] V.S. Varadarajan. Groups of automorphisms of Borel spaces. *Trans. Amer. Math. Soc.*, **109** (1963), 191–220. 38

[Wa] V.M. Wagh. A descriptive version of Ambrose's representation theorem for flows. *Proc. Indian Acad. Sci. Math. Sci.*, **98** (1988), 101–108. 25

[We1] B. Weiss. Measurable dynamics. *Contemp. Math.*, **26** (1984), 395–421. 36, 40, 48, 59, 69

[We2] B. Weiss. Countable generators in dynamics – universal minimal models. *Contemp. Math.*, **94** (1989), 321–326. 117

[W1] J. Williams. Universal countable Borel quasi-orders. *J. Symb. Logic*, **79** (2014), 928–954. 115, 116

[W2] J. Williams. Isomorphism of finitely generated solvable groups is weakly universal. *J. Pure Appl. Algebra*, **219**(5) (2015), 1639–1644. 118

[Z1] R.J. Zimmer. Hyperfinite factors and amenable ergodic actions. *Invent. Math.*, **41** (1977), 23–31. 88

[Z2] R.J. Zimmer. *Ergodic Theory and Semisimple Groups*. Birkhäuser, 1984. xii, 52, 55, 88

List of Notation

Subject Index

Printed in the United States
by Baker & Taylor Publisher Services